This notebook belongs to:

Name: _ _ _ _ _ _ _ _ _ _

Adress: _ _ _ _ _ _ _ _ _

Phone: _ _ _ _ _ _ _ _ _

Thank you for choosing BOOKSZOOKA Notebooks!

   As a small family company, we value your opinions and advice to create better products. We hope to give all of our clients the best service possible. Please let us know about your experience working with us. Your feedback is both appreciated and helpful.

   Also, we would like to hear from you and see your work:

 @BooksZooka

 BooksZooka

 hello@bookszooka.com

Multiplication table

0	1	2	3	4	5	6	7	8	9	10
1	1	2	3	4	5	6	7	8	9	10
2	2	4	6	8	10	12	14	16	18	20
3	3	6	9	12	15	18	21	24	27	30
4	4	8	12	16	20	24	28	32	36	40
5	5	10	15	20	25	30	35	40	45	50
6	6	12	18	24	30	36	42	48	54	60
7	7	14	21	28	35	42	49	56	63	70
8	8	16	24	32	40	48	56	64	72	80
9	9	18	27	36	45	54	63	72	81	90
10	10	20	30	40	50	60	70	80	90	100

Math conversion chart - Lengths

Standard conversion

1 foot (ft)	=	12 inches (in)
1 yard (yd)	=	3 feet (ft)
1 yard (yd)	=	36 inches (in)
1 mile (mi)	=	1760 yards (yd)

Metric conversion

1 centimeter (cm)	=	10 millimeters (mm)
1 meter (m)	=	100 centimeters (cm)
1 kilometer (km)	=	1000 meters (m)

Standard to Metric conversion

1 inch (in)	=	2.54 centimeters (cm)
1 foot (ft)	=	30.48 centimeters (cm)
1 yard (yd)	=	91.44 centimeters (cm)
1 yard (yd)	=	0.9144 meters (m)
1 mile (mi)	=	1609.344 meters (m)
1 mile (mi)	=	1.609344 kilometers (km)

Metric to Standard conversion

1 millimeter (mm)	=	0.03937 inches (in)
1 centimeter (cm)	=	0.39370 inches (in)
1 meter (m)	=	39.37008 inches (in)
1 meter (m)	=	3.28084 feet (ft)
1 meter (m)	=	1.09361 yards (yd)
1 kilometer (km)	=	1093.6133 yards (yd)
1 kilometer (km)	=	0.62137 miles (mi)

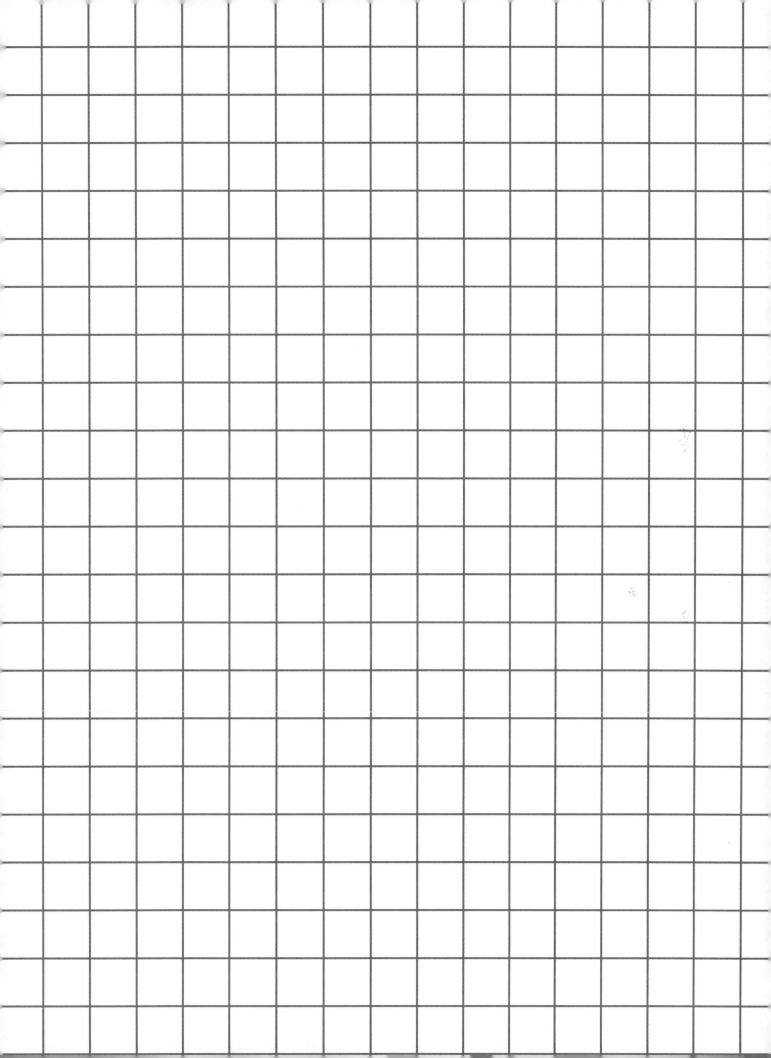

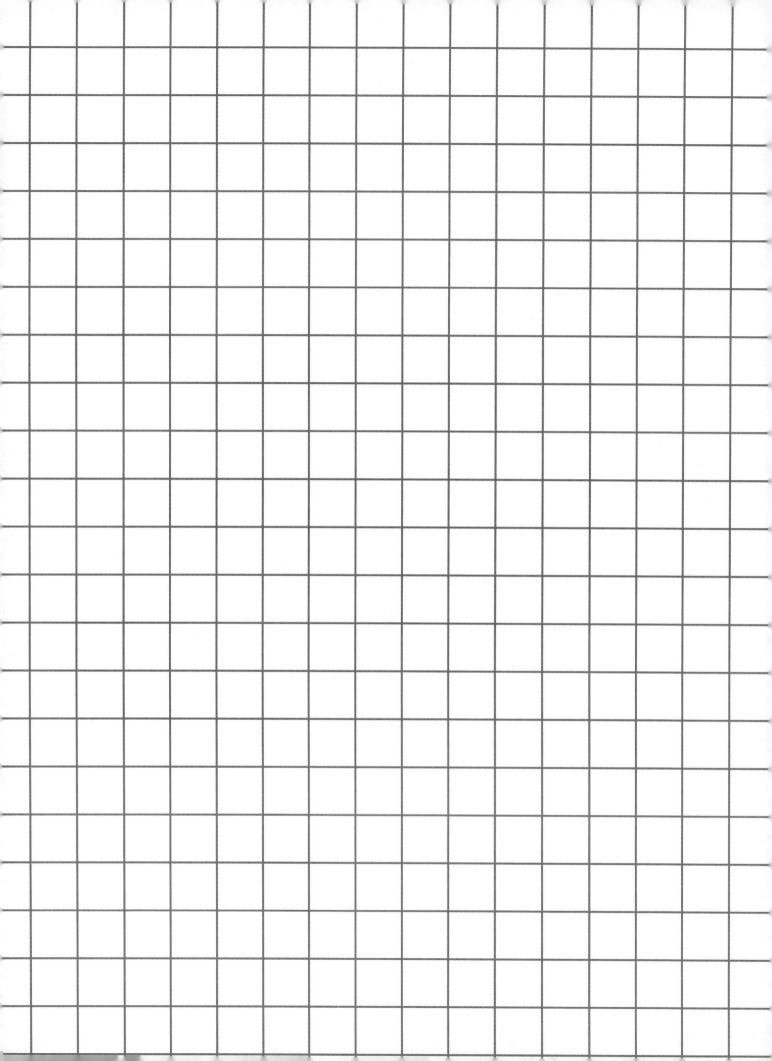

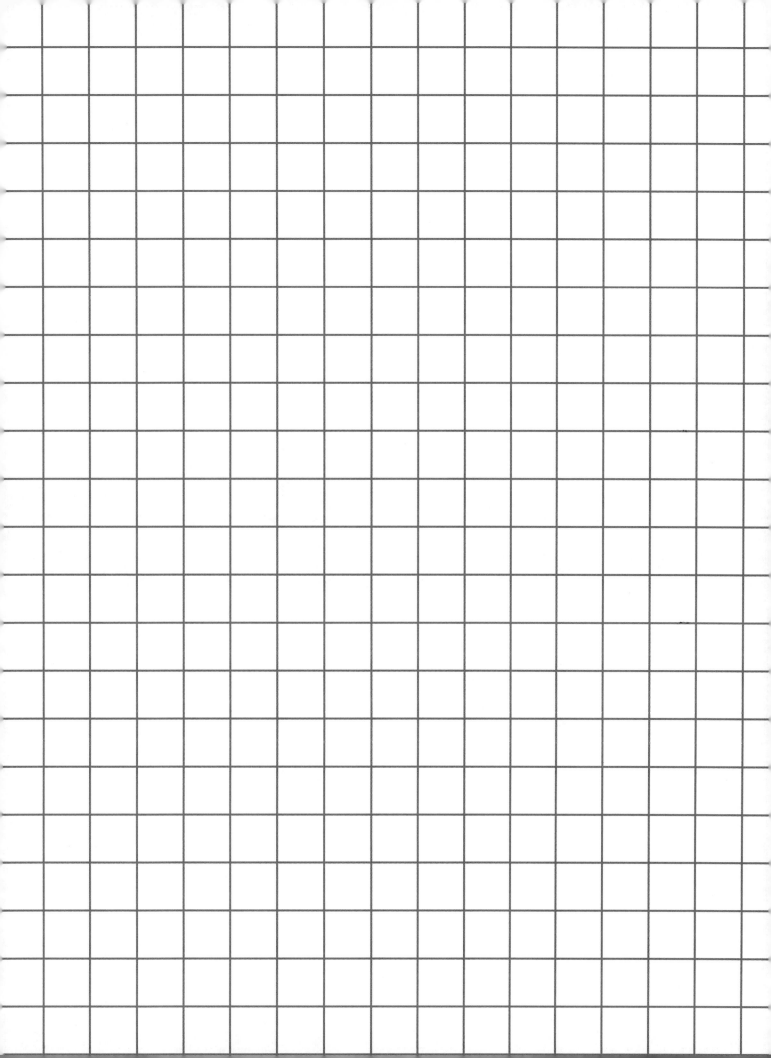

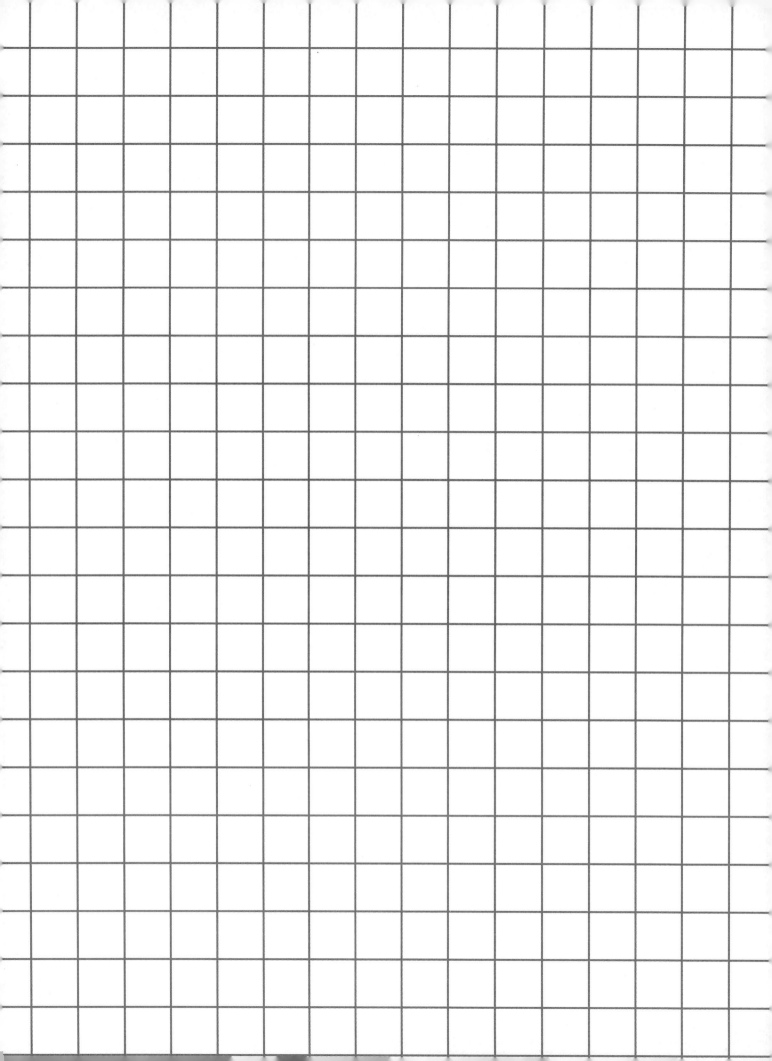

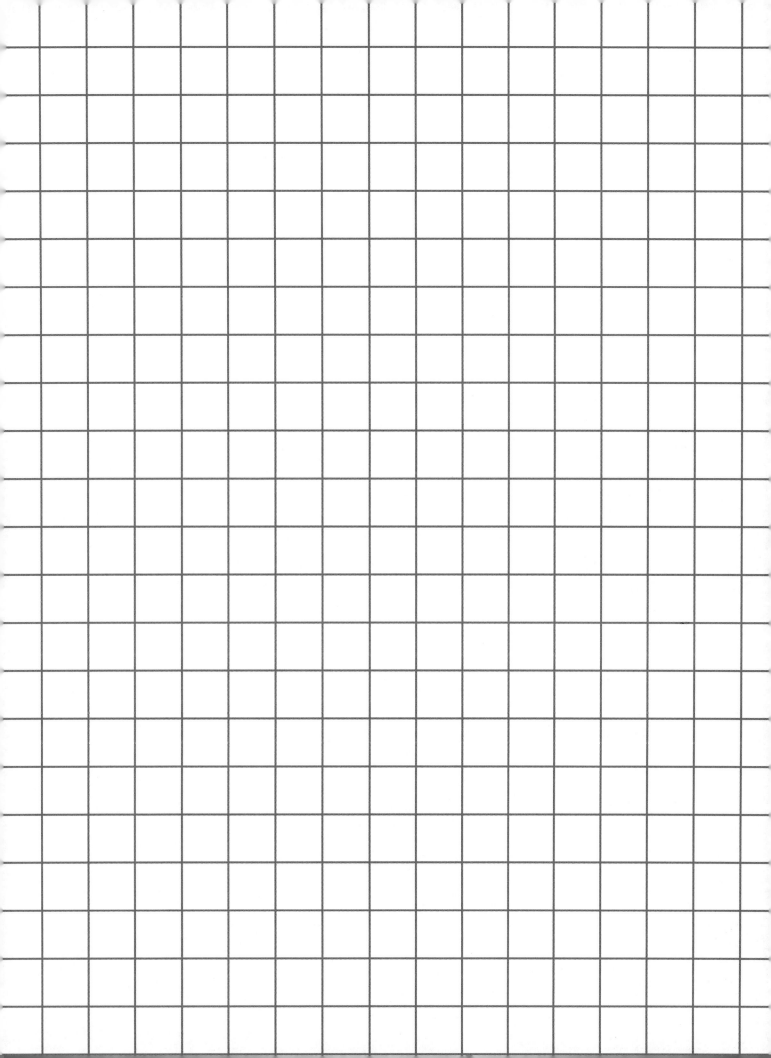

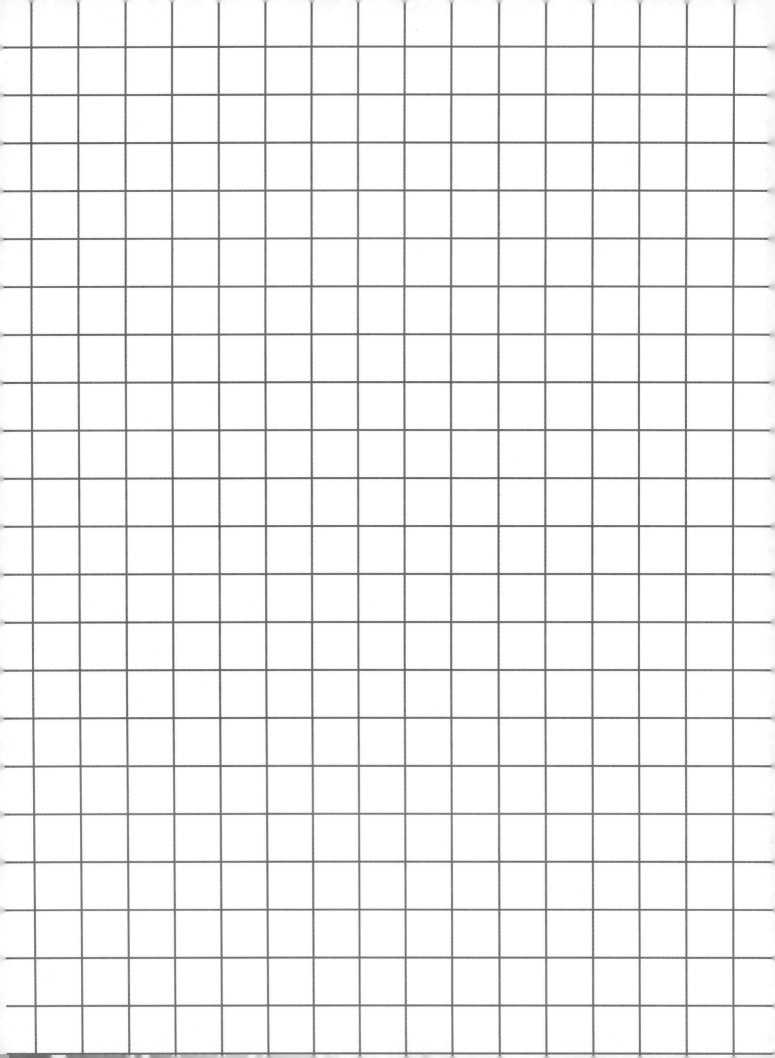

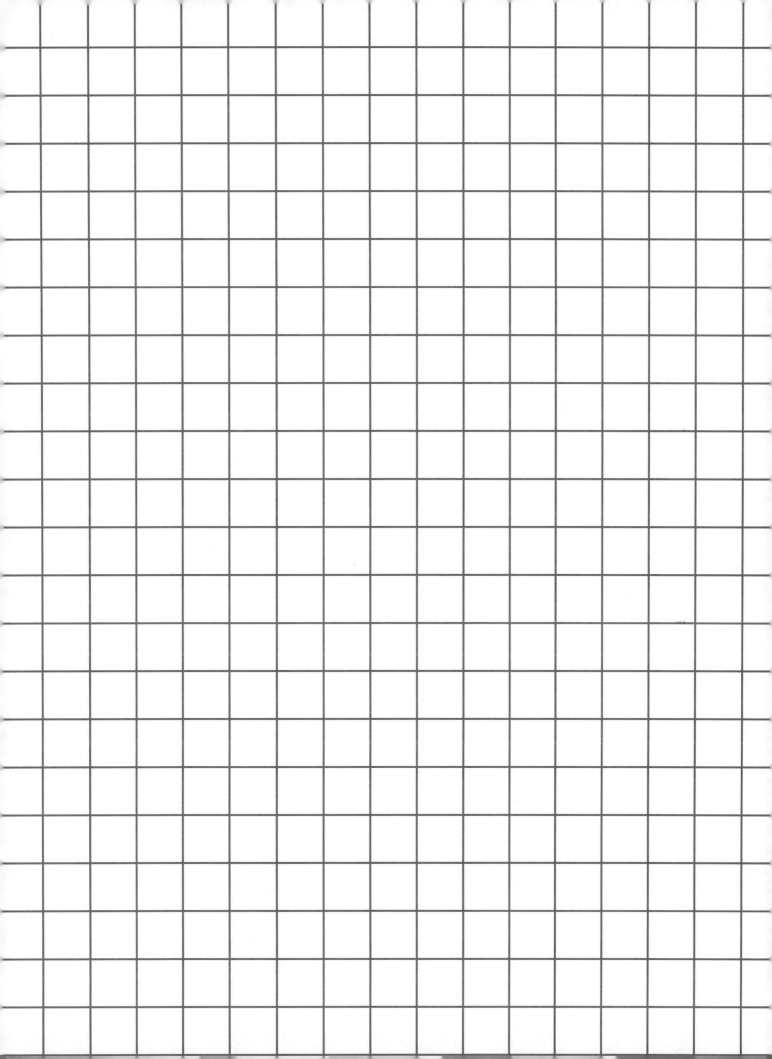

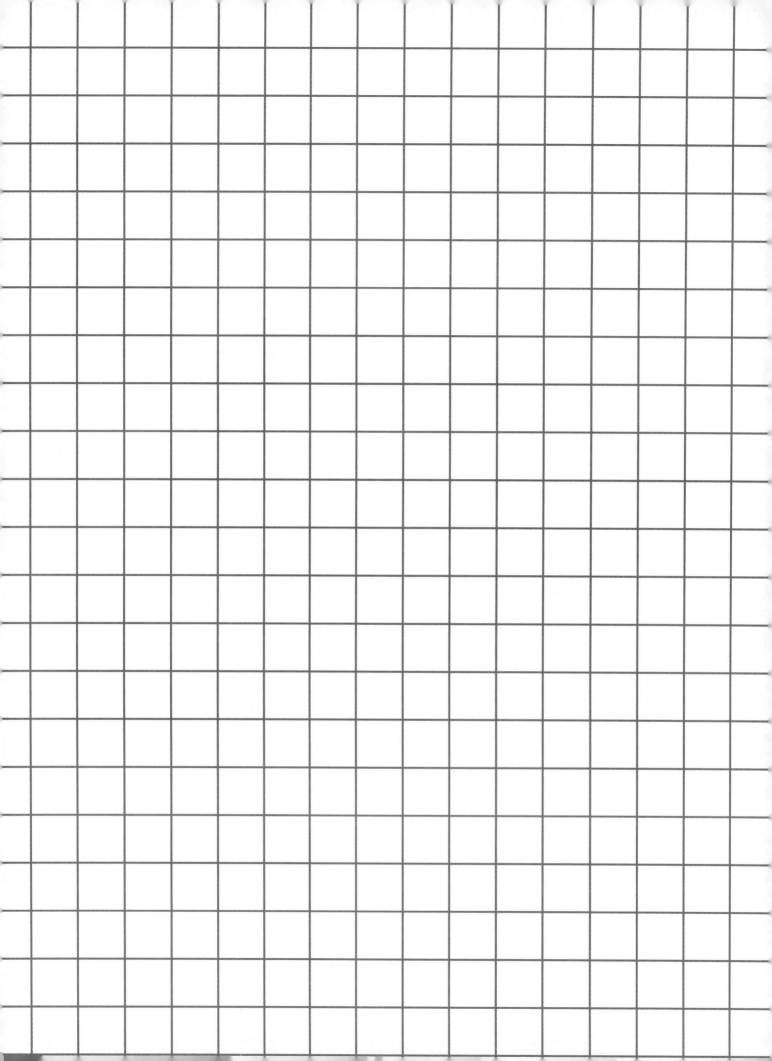

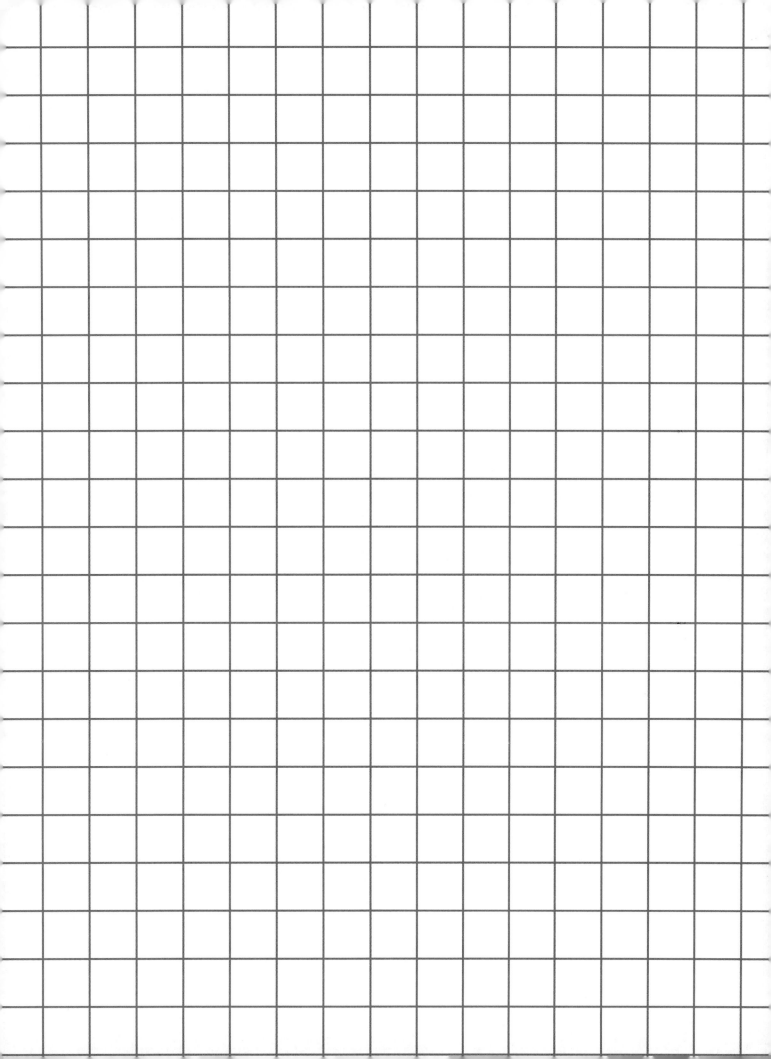

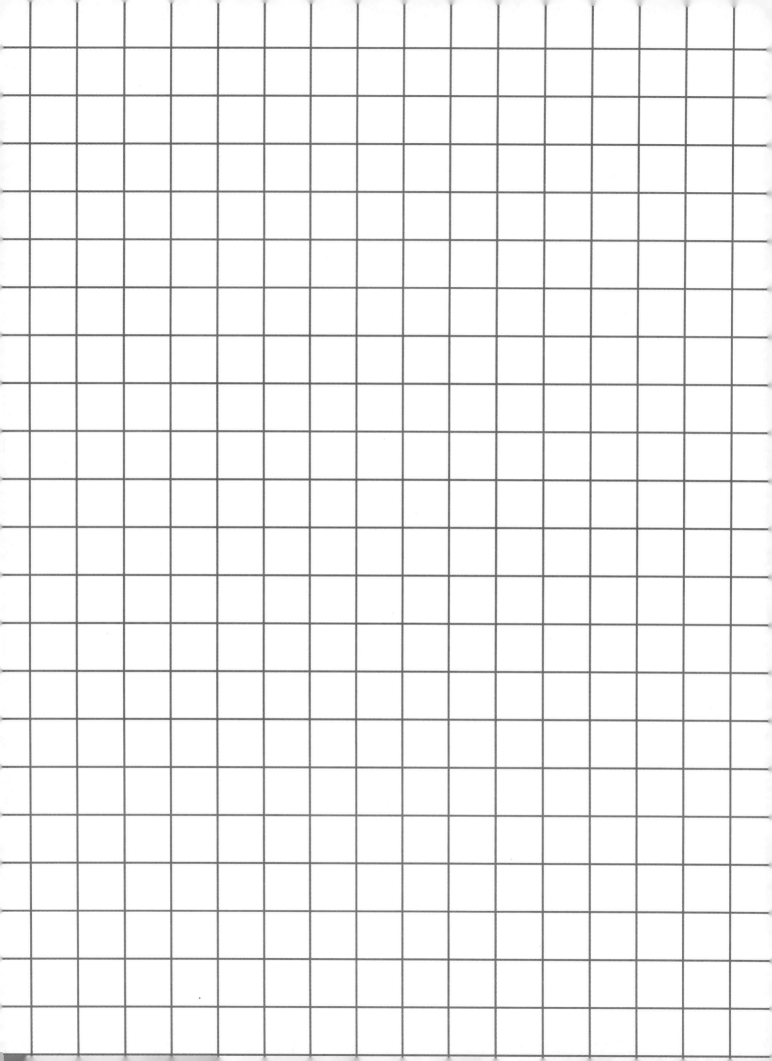

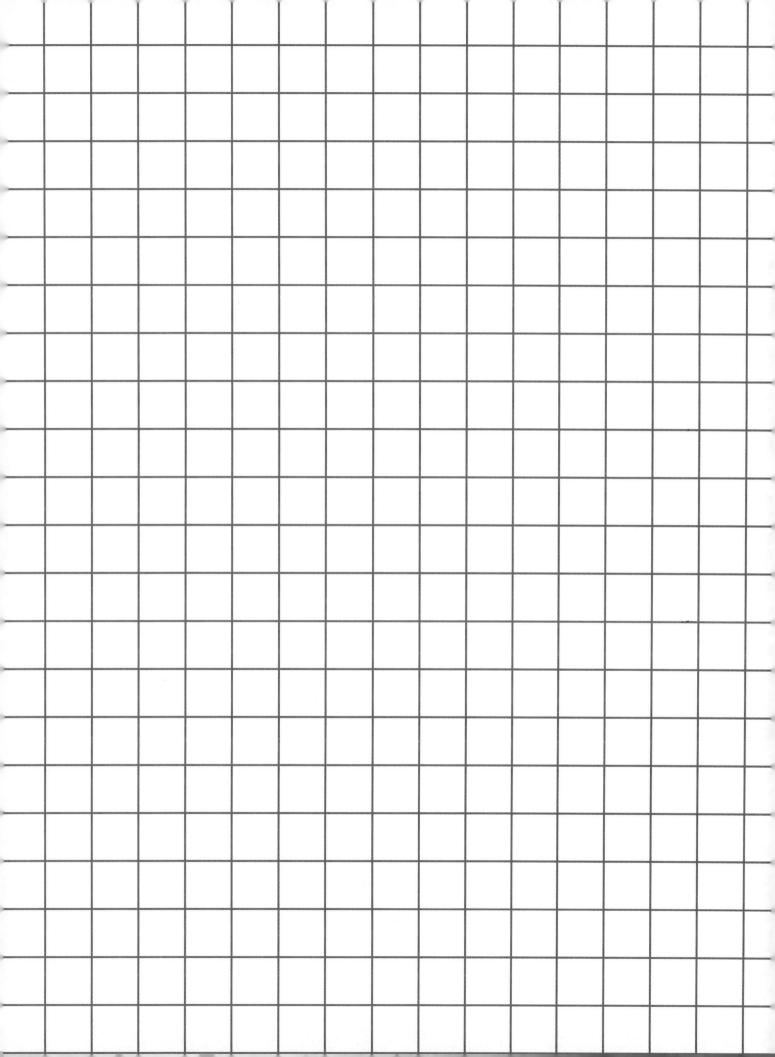

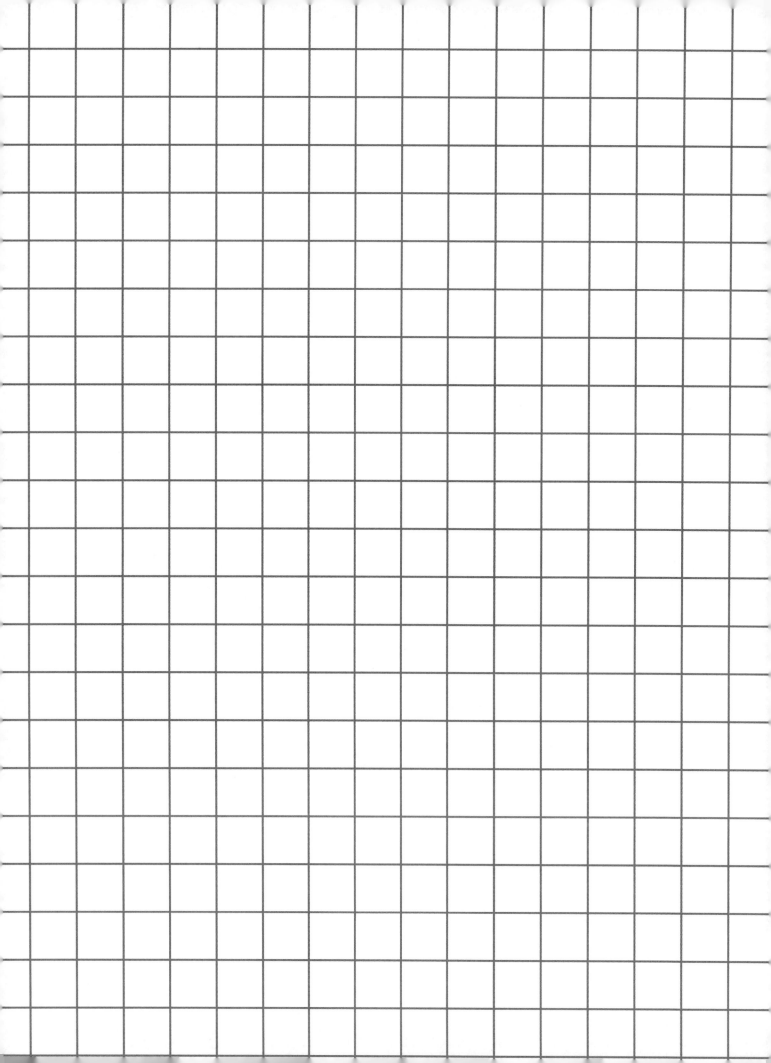

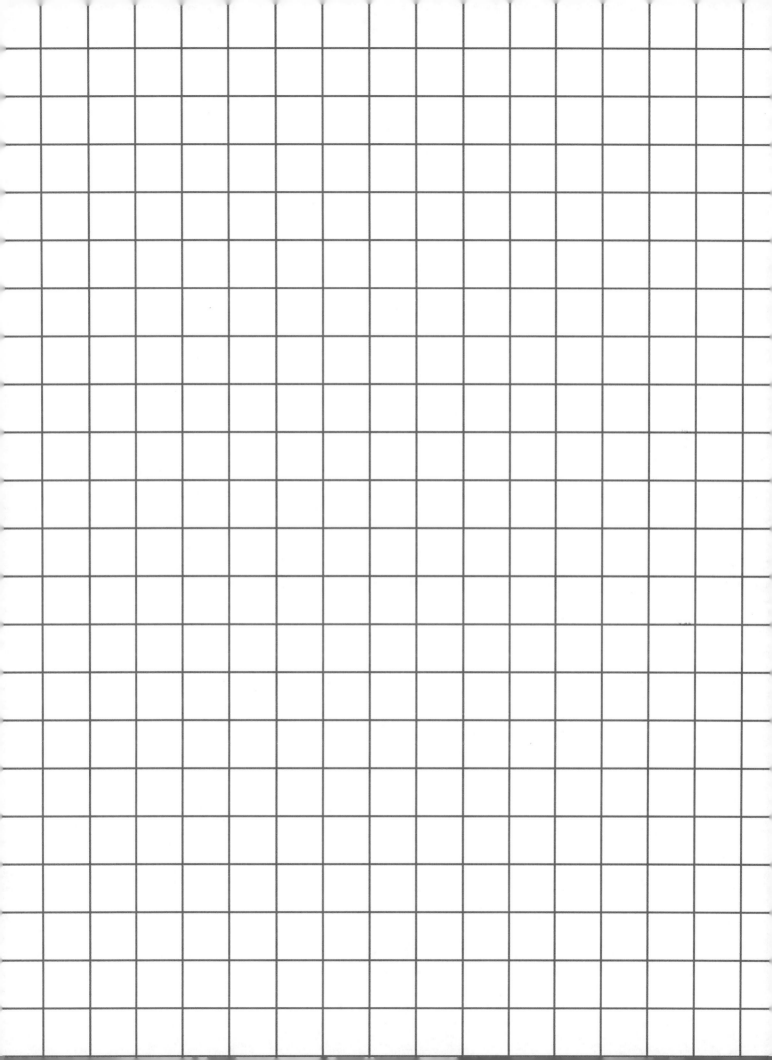

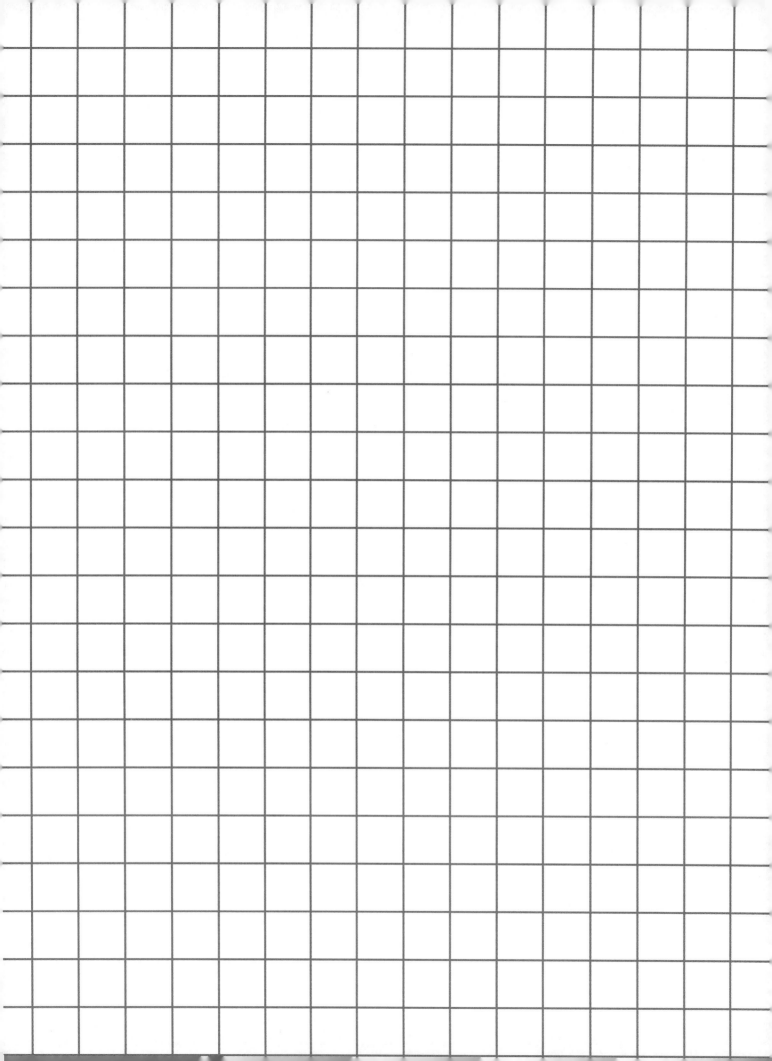

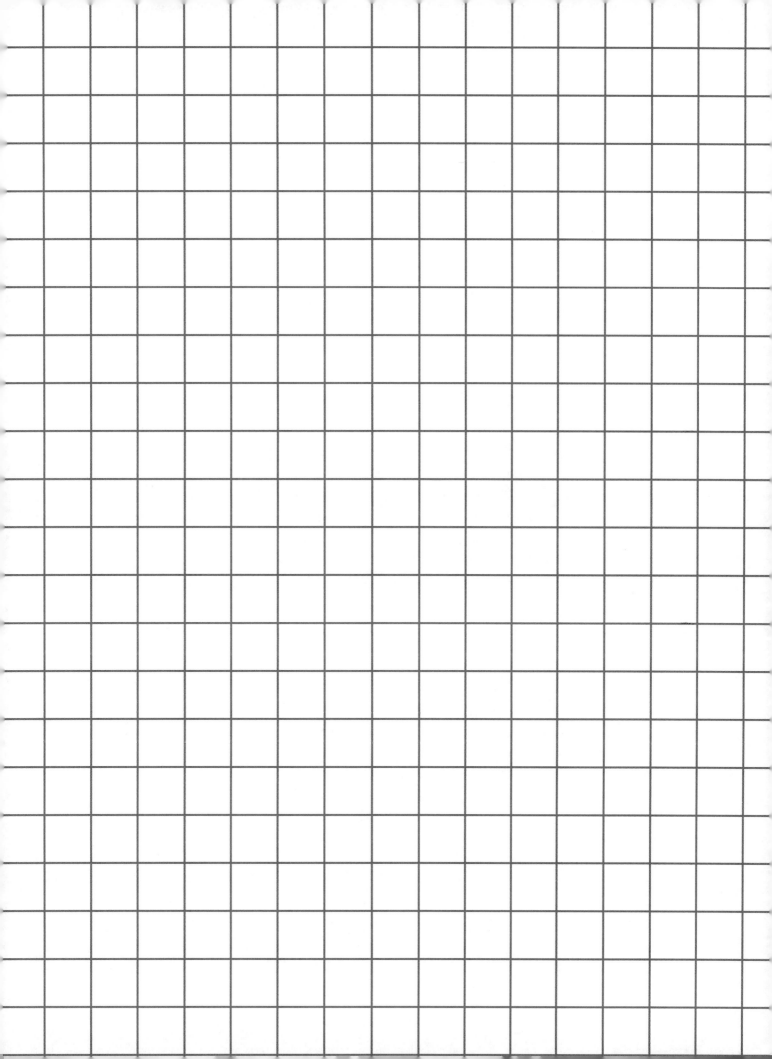

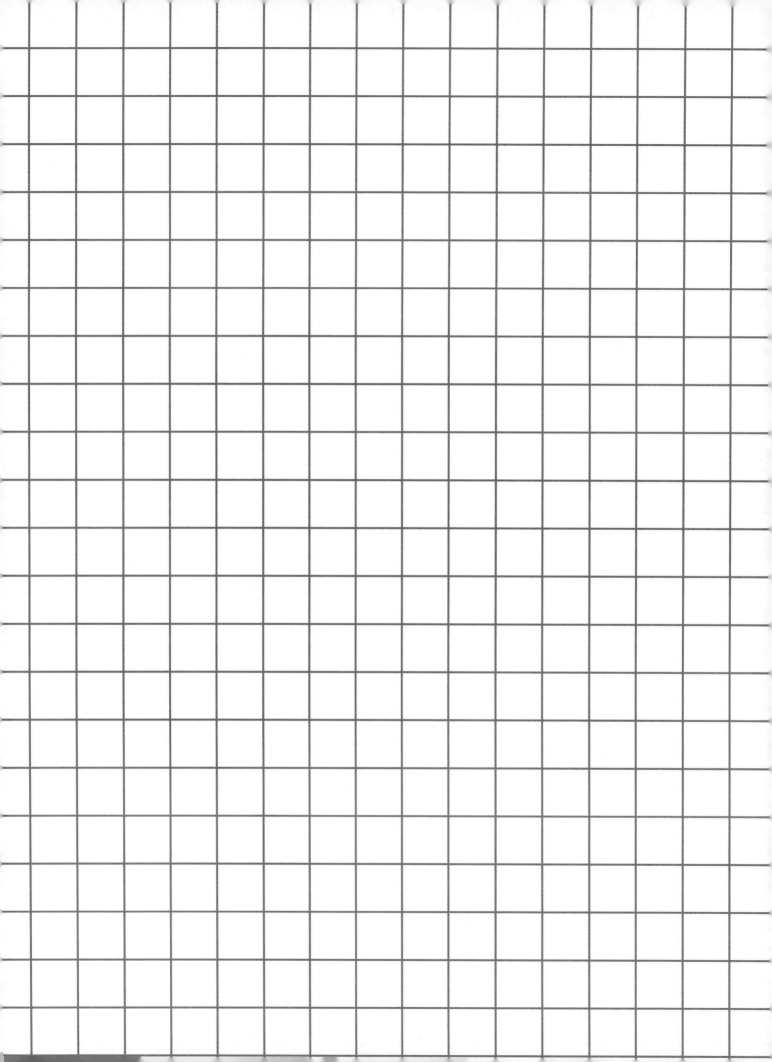

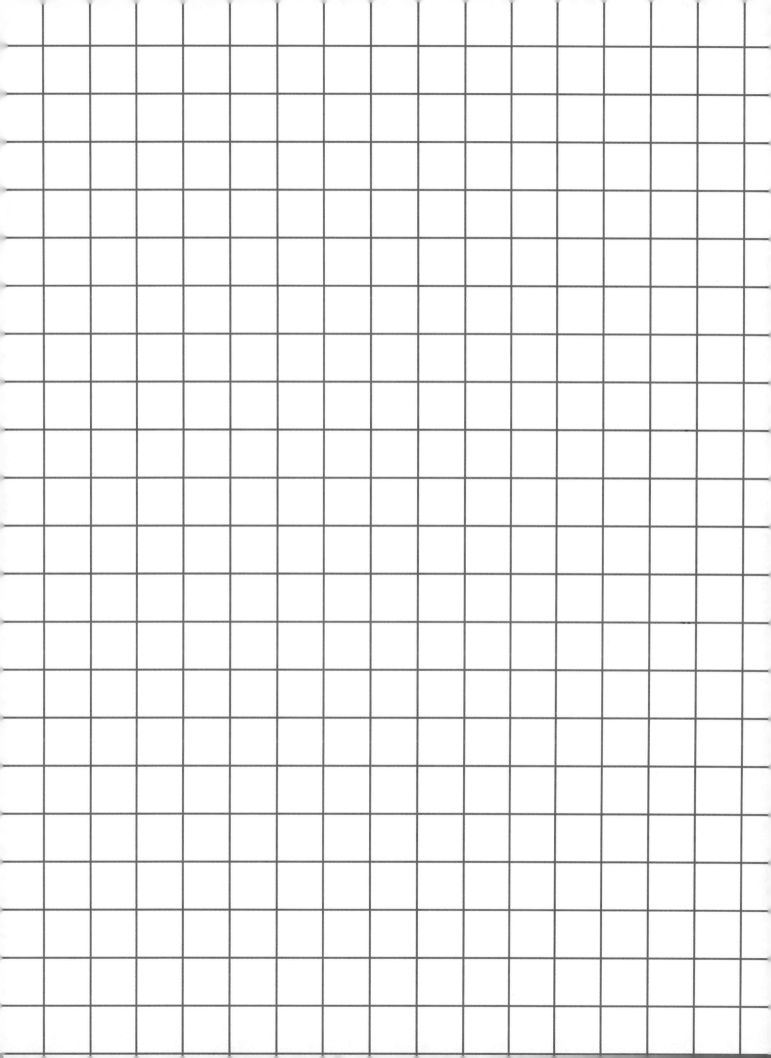

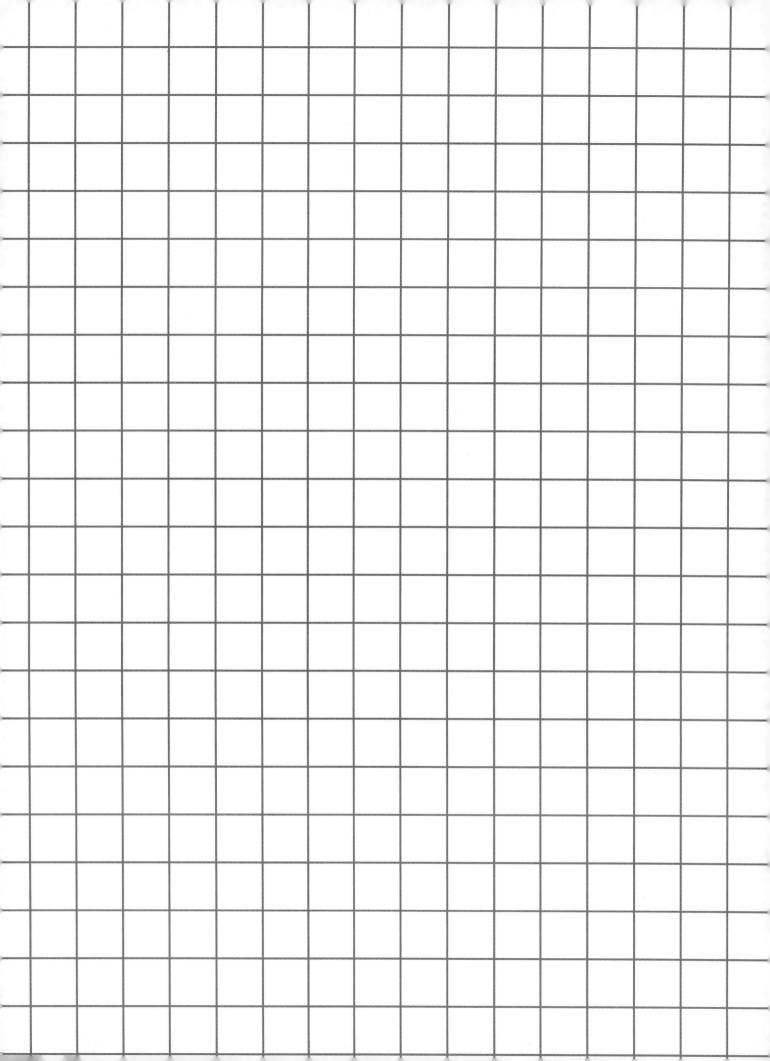

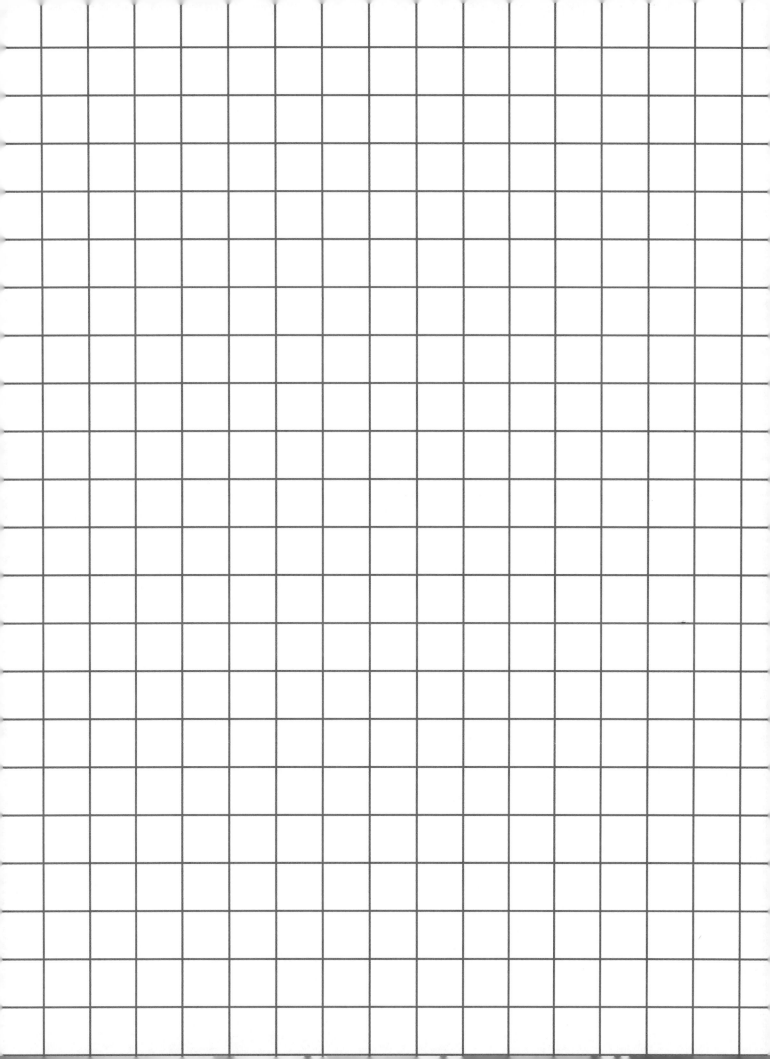

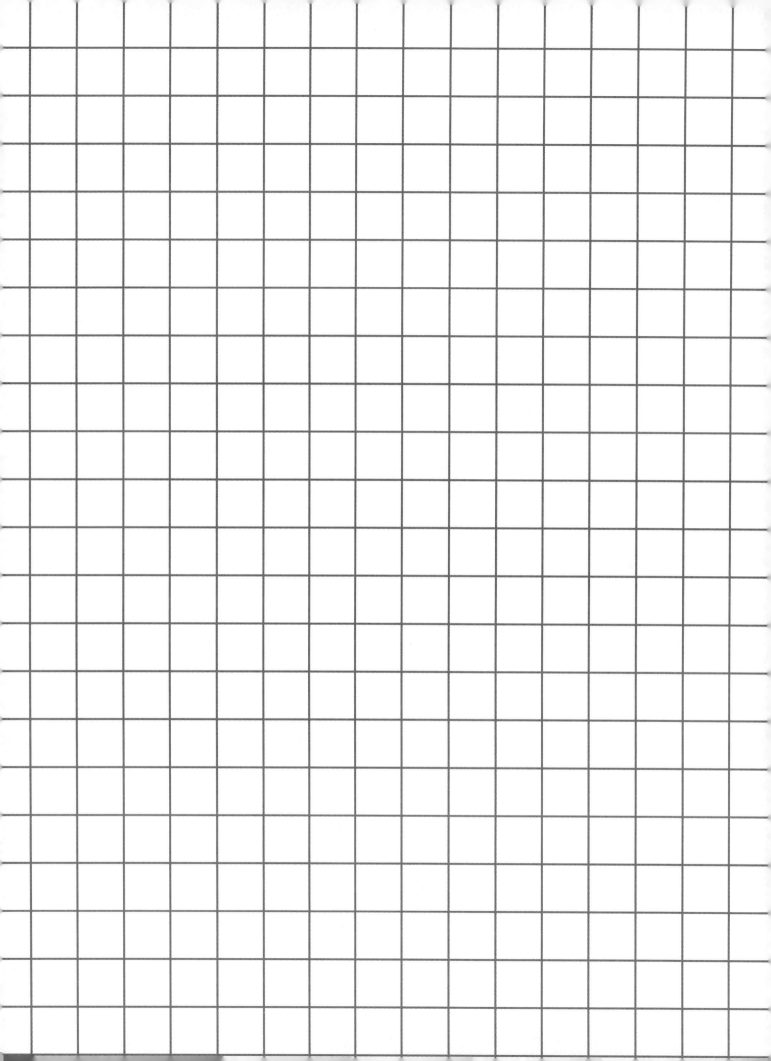

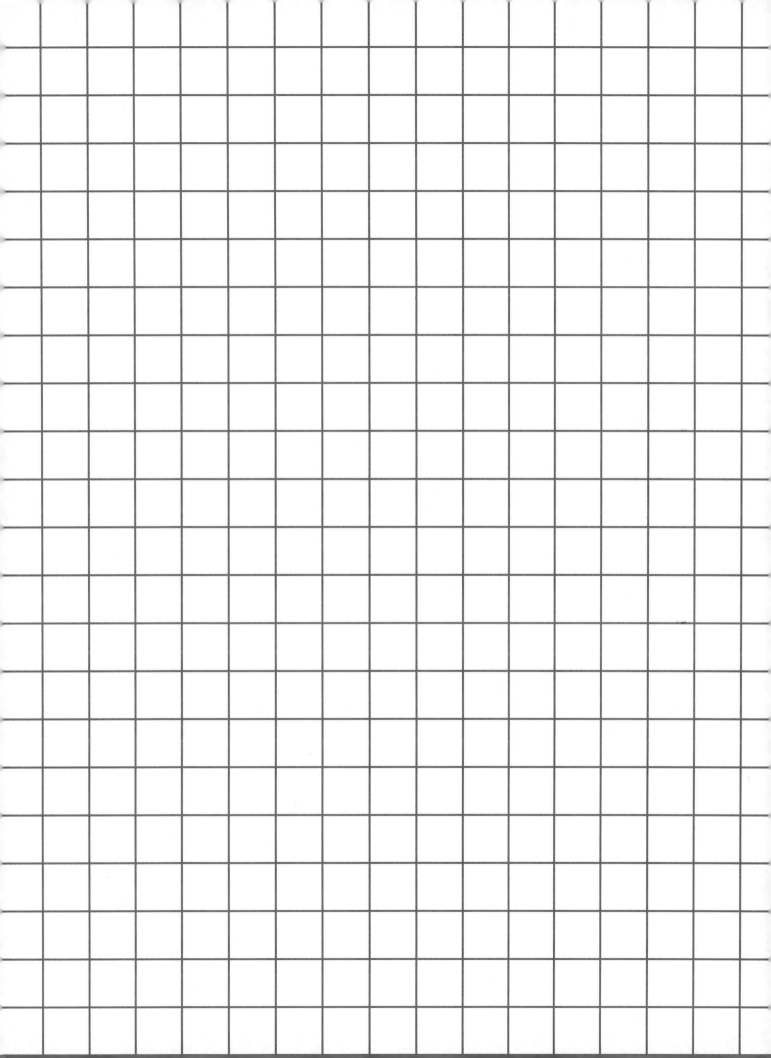

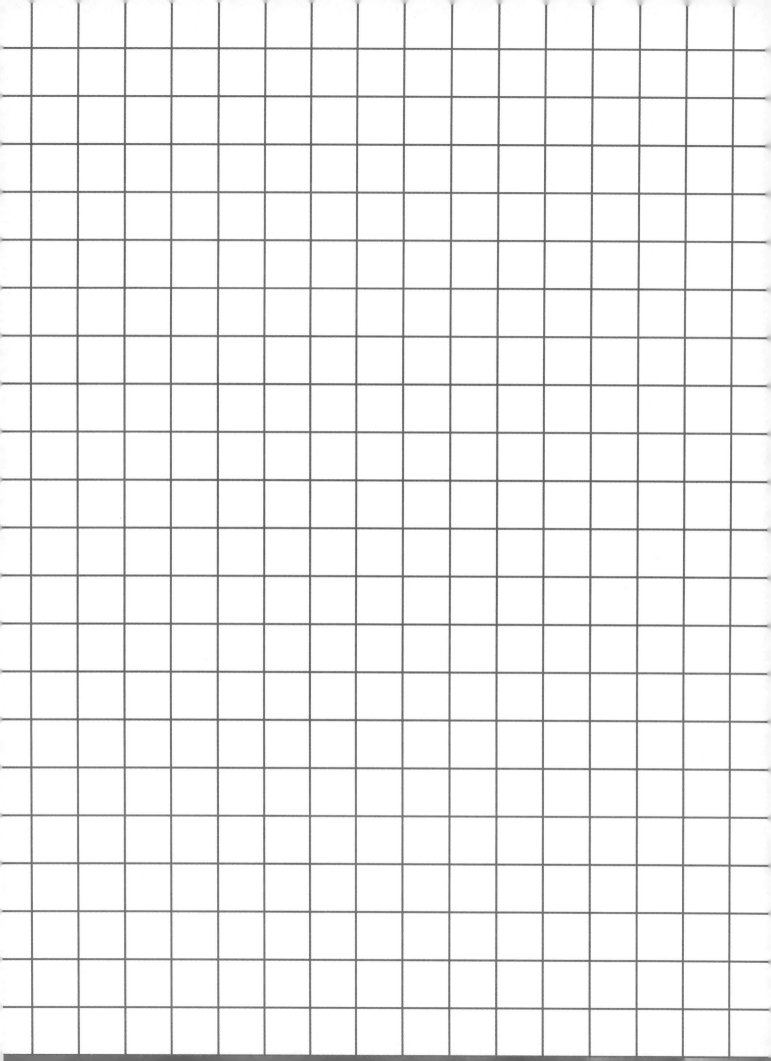

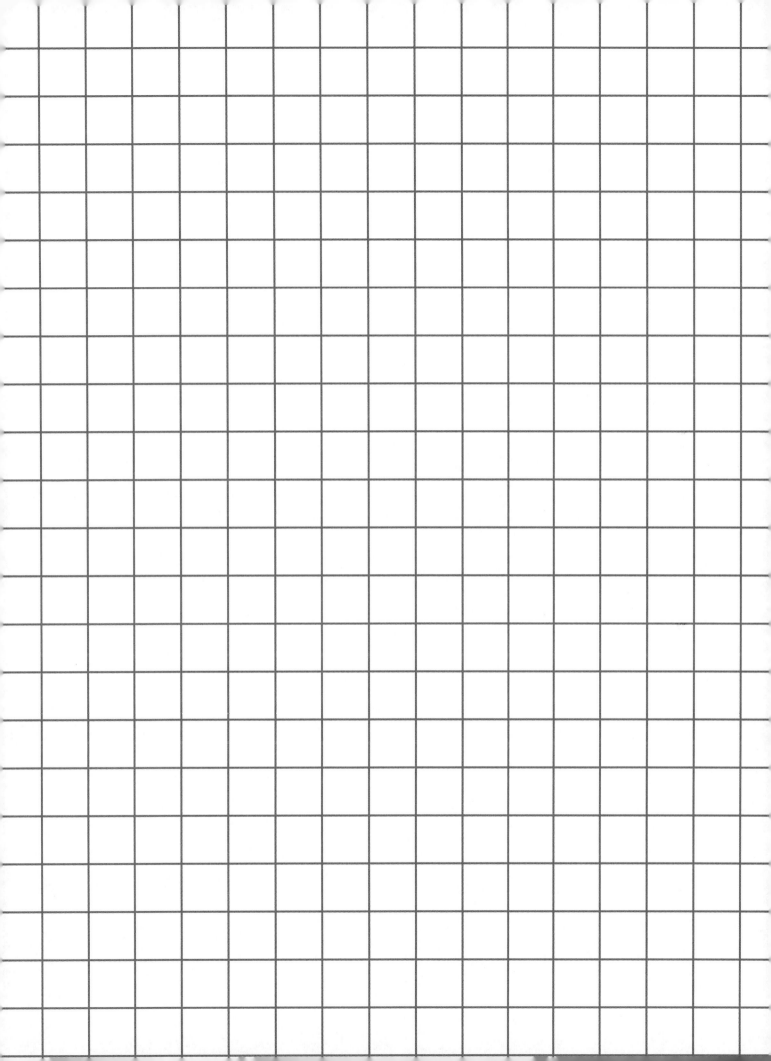

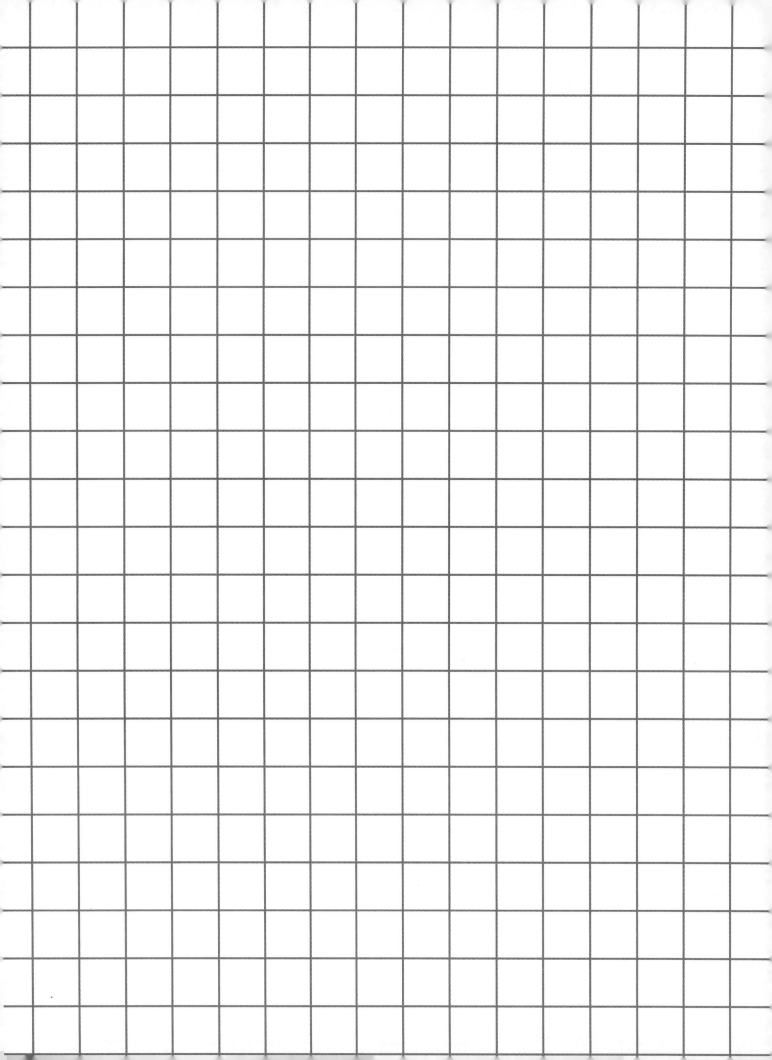

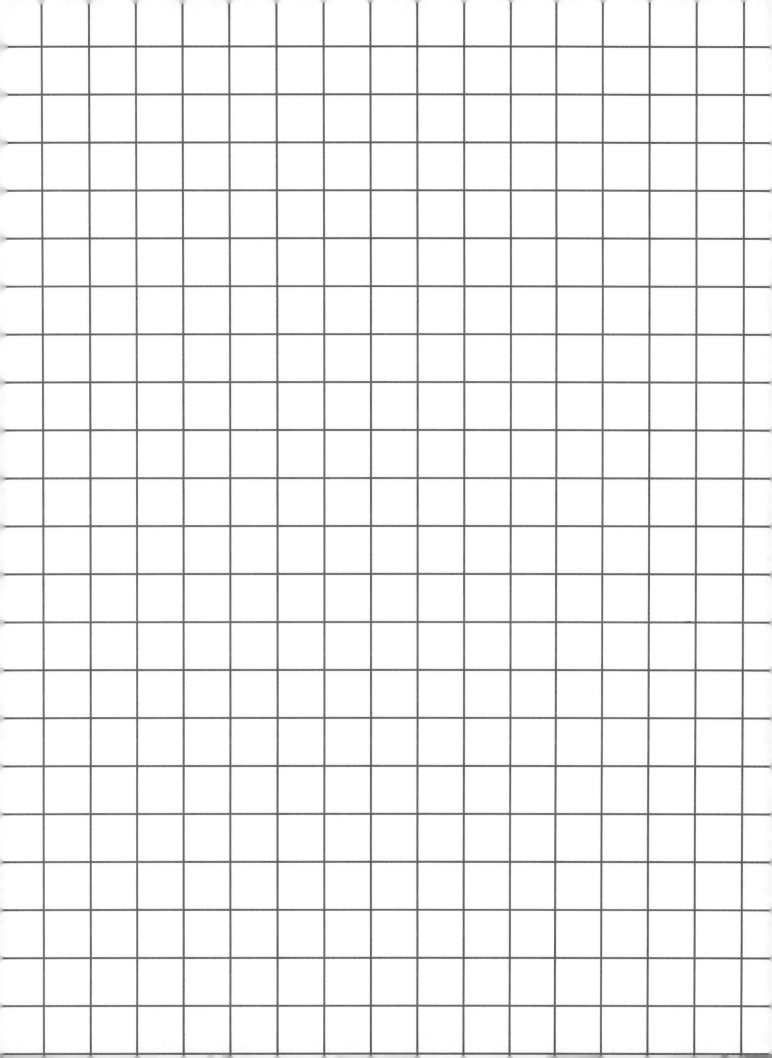

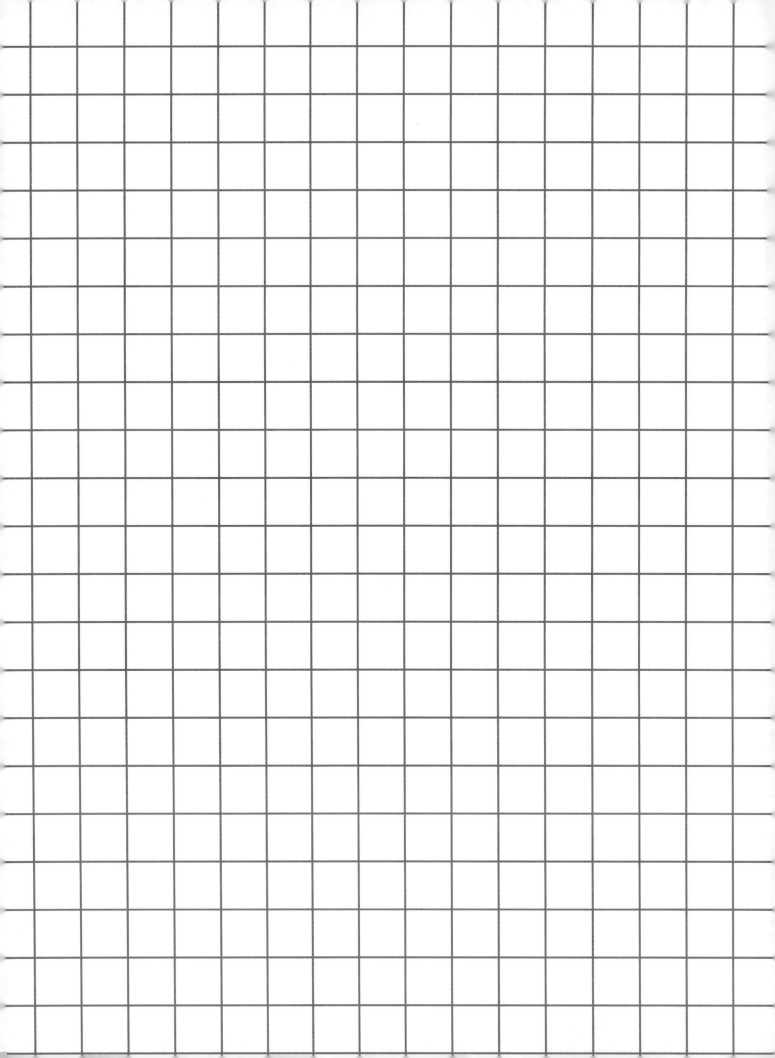

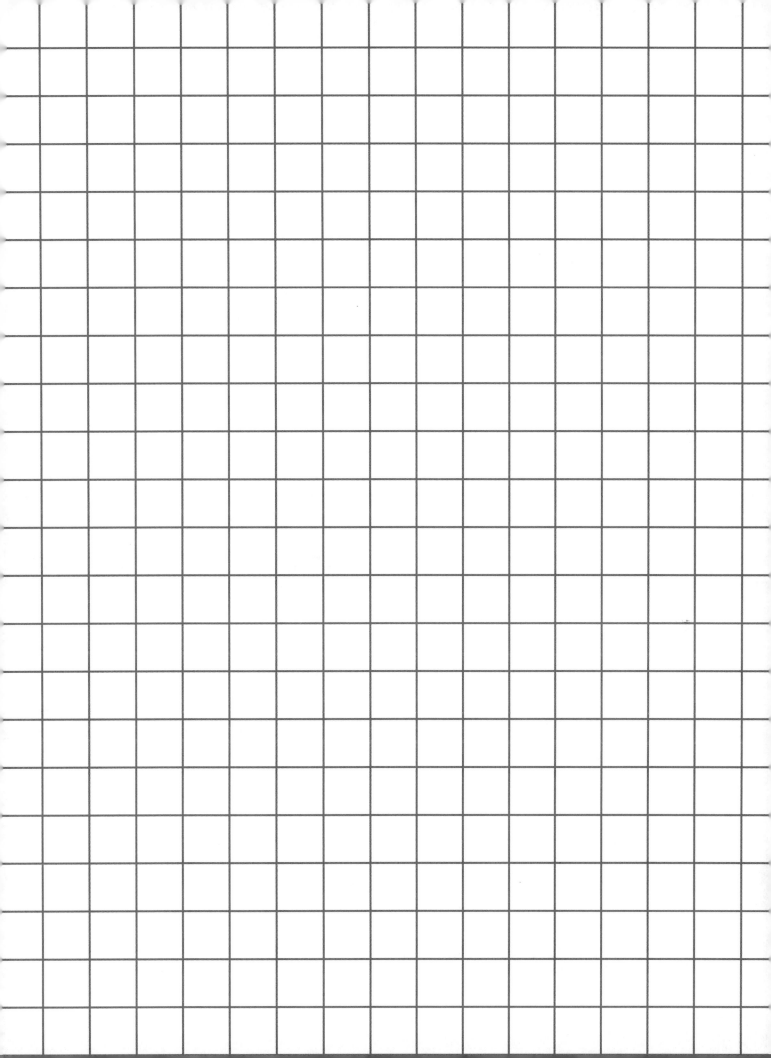

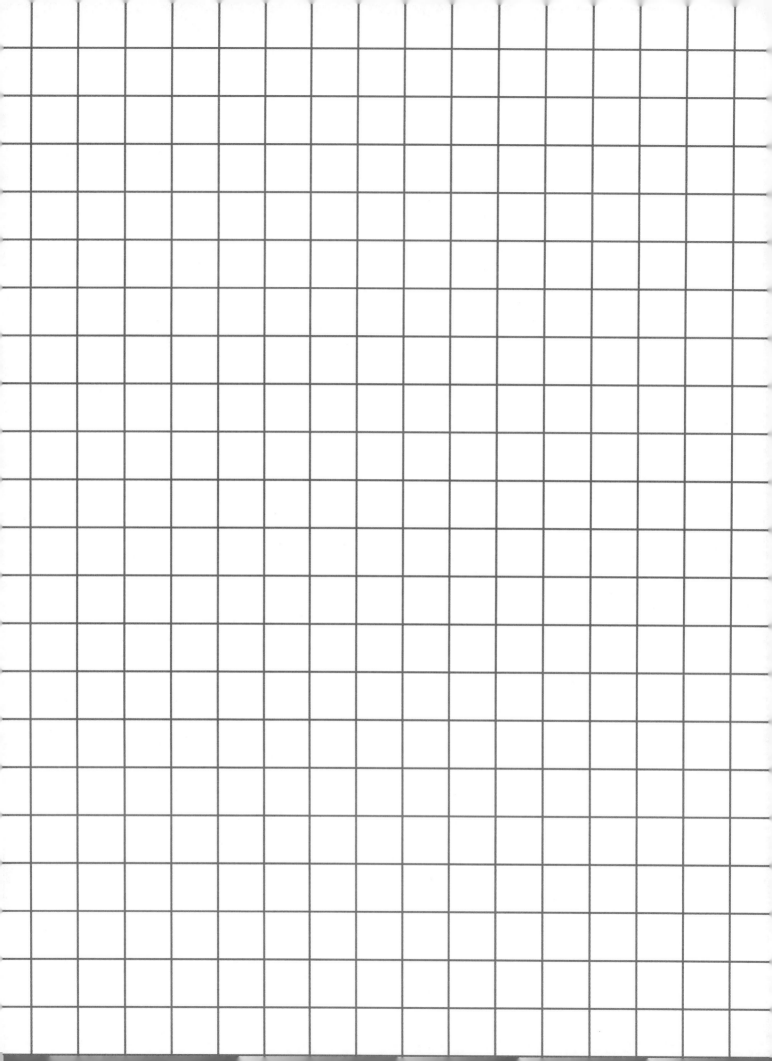

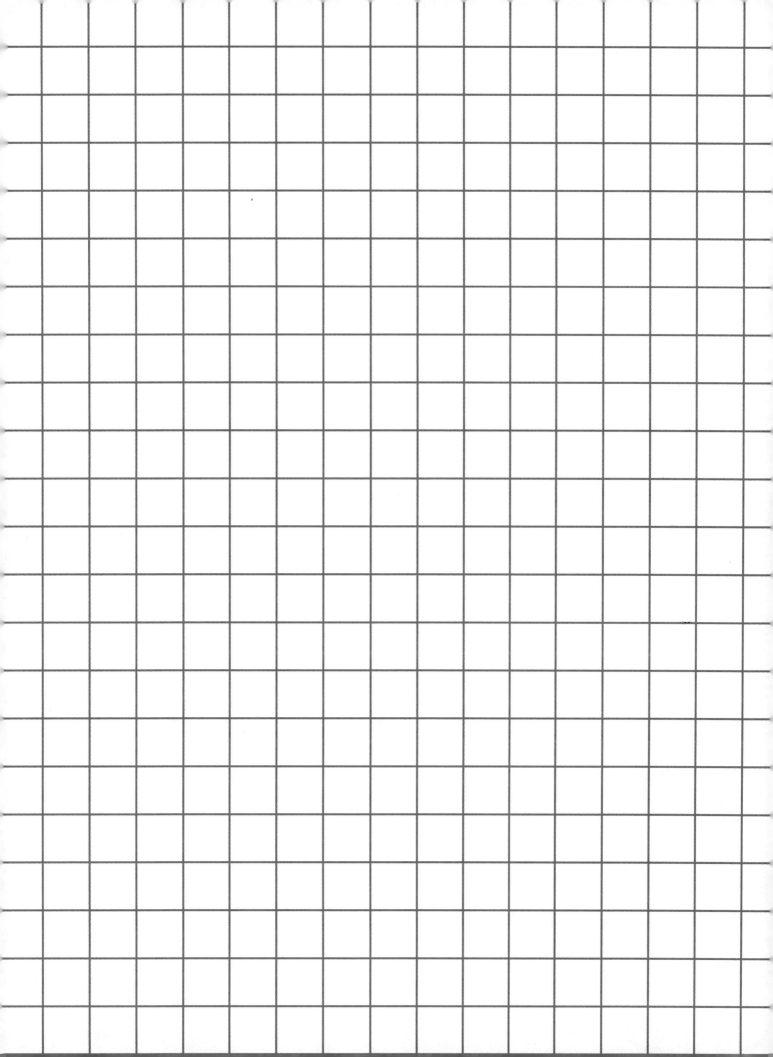

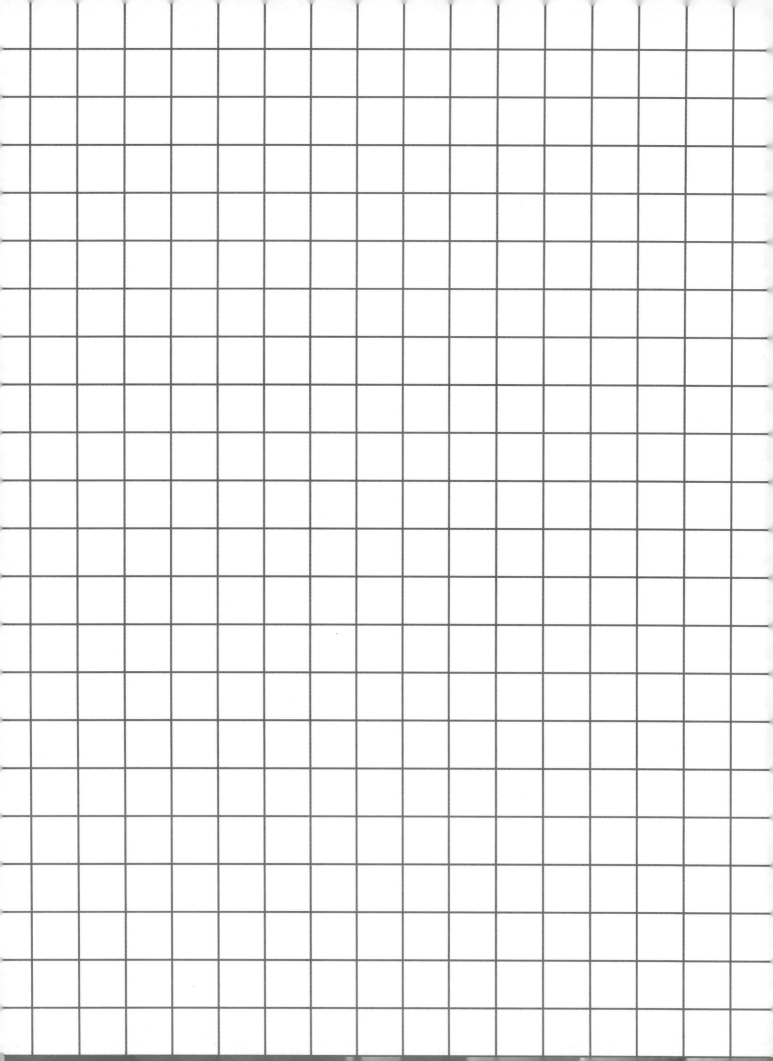

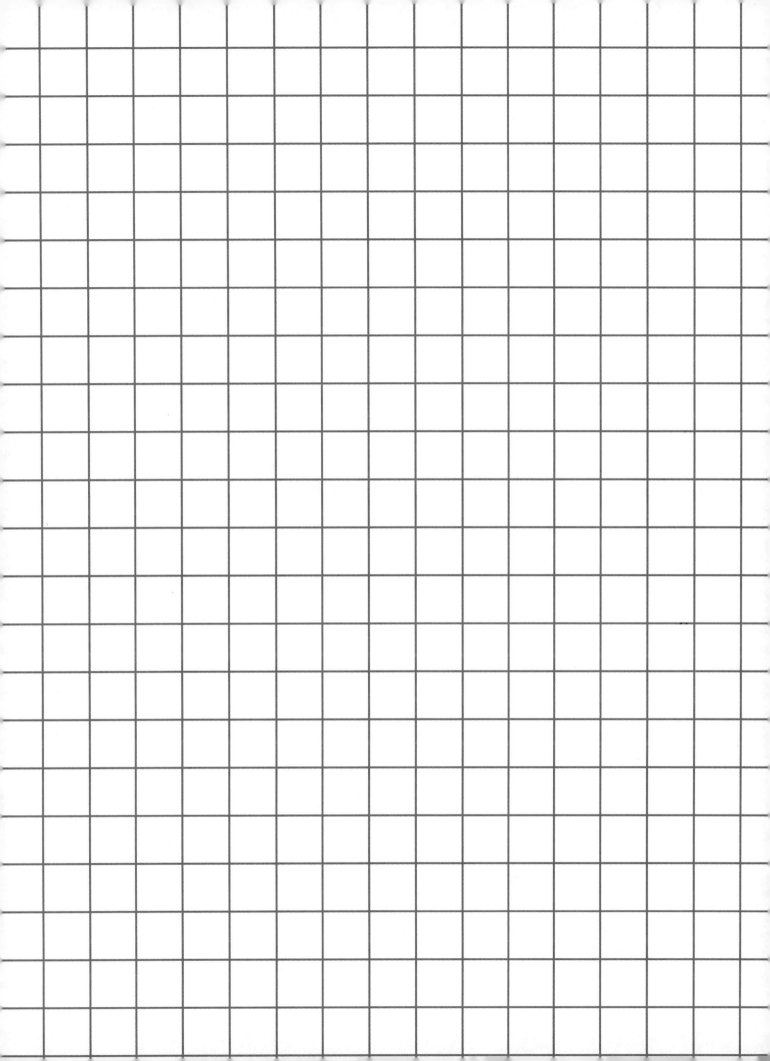

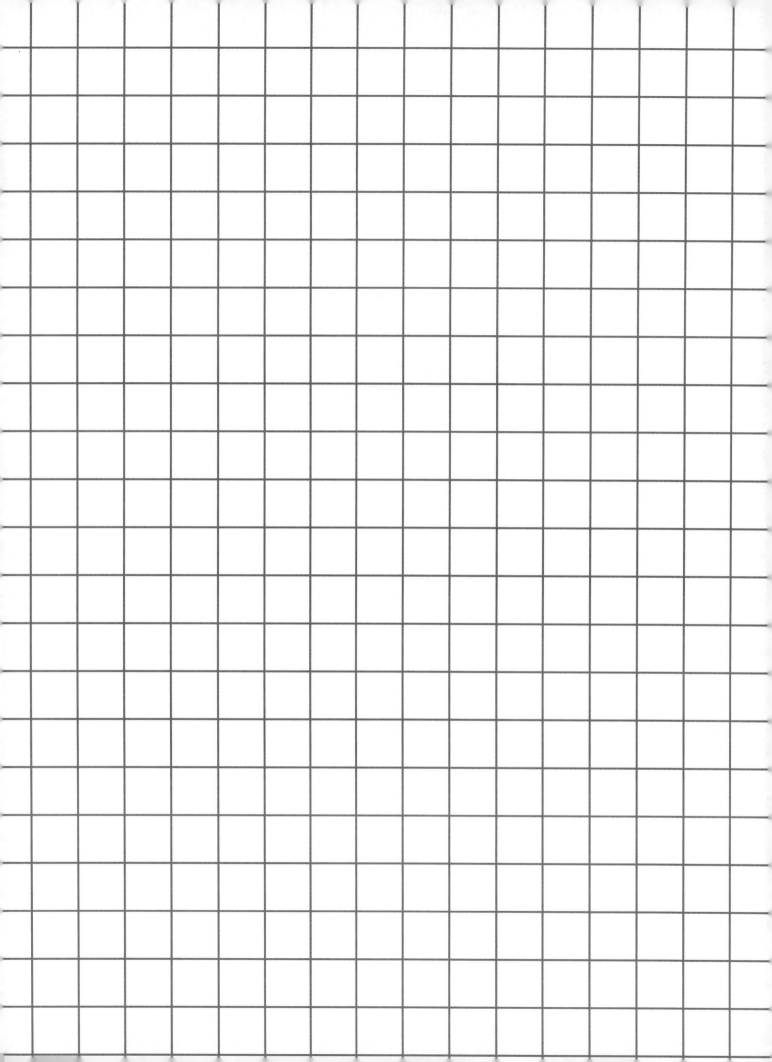

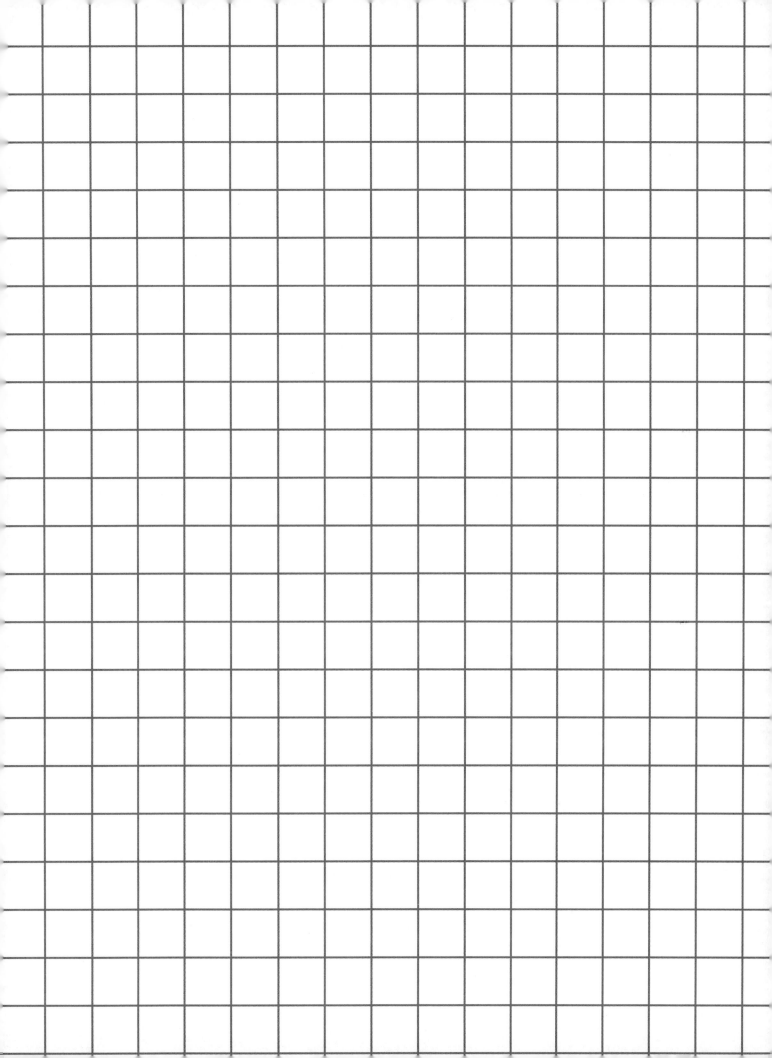

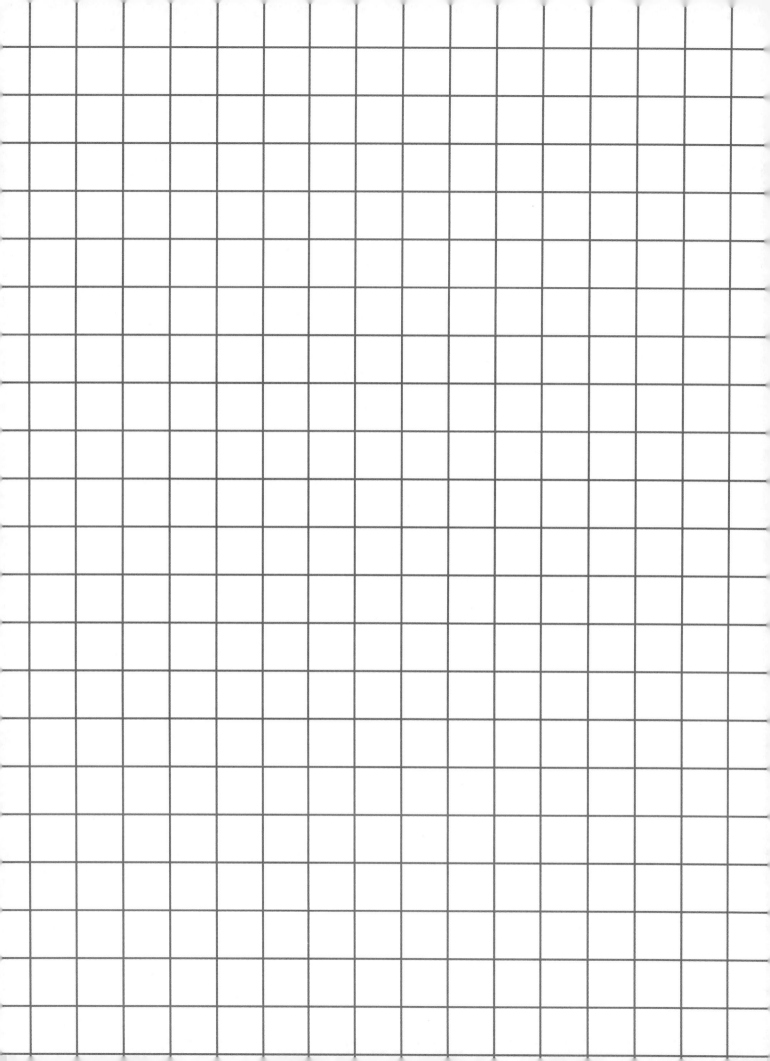

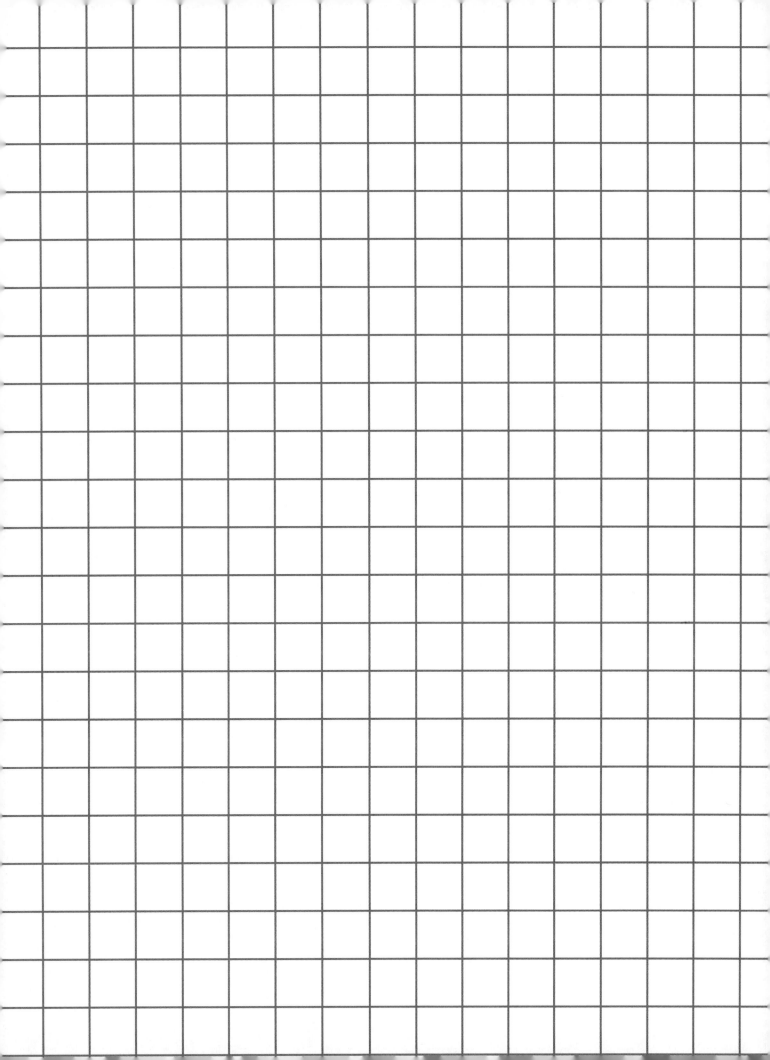

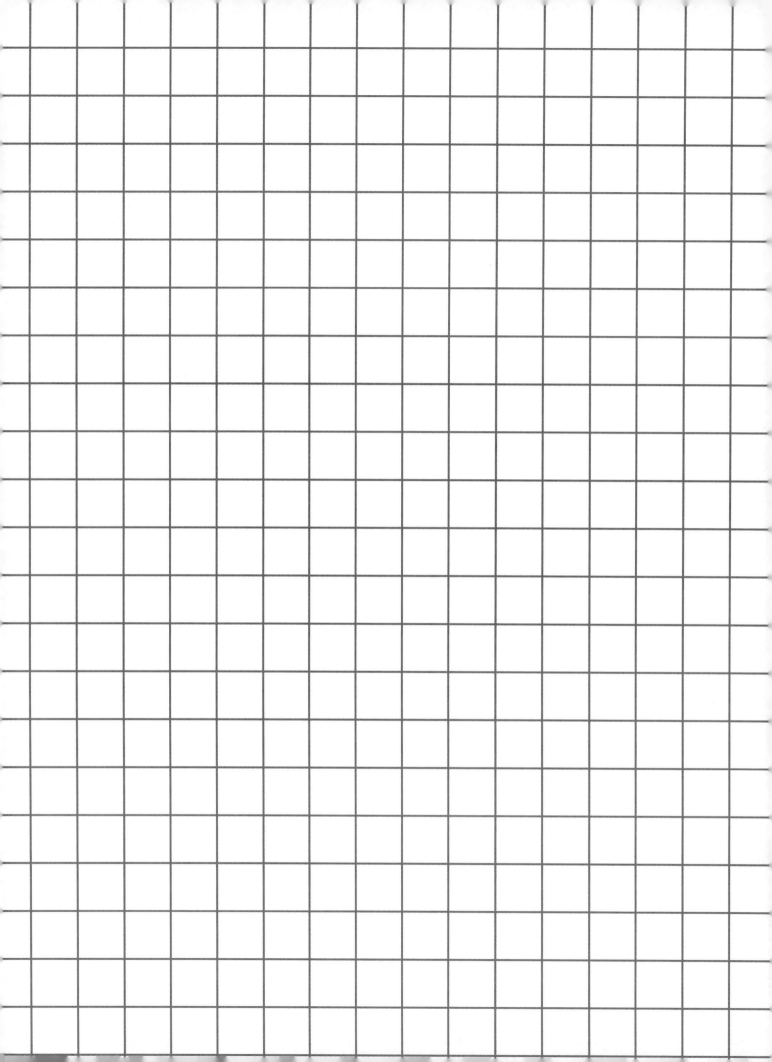

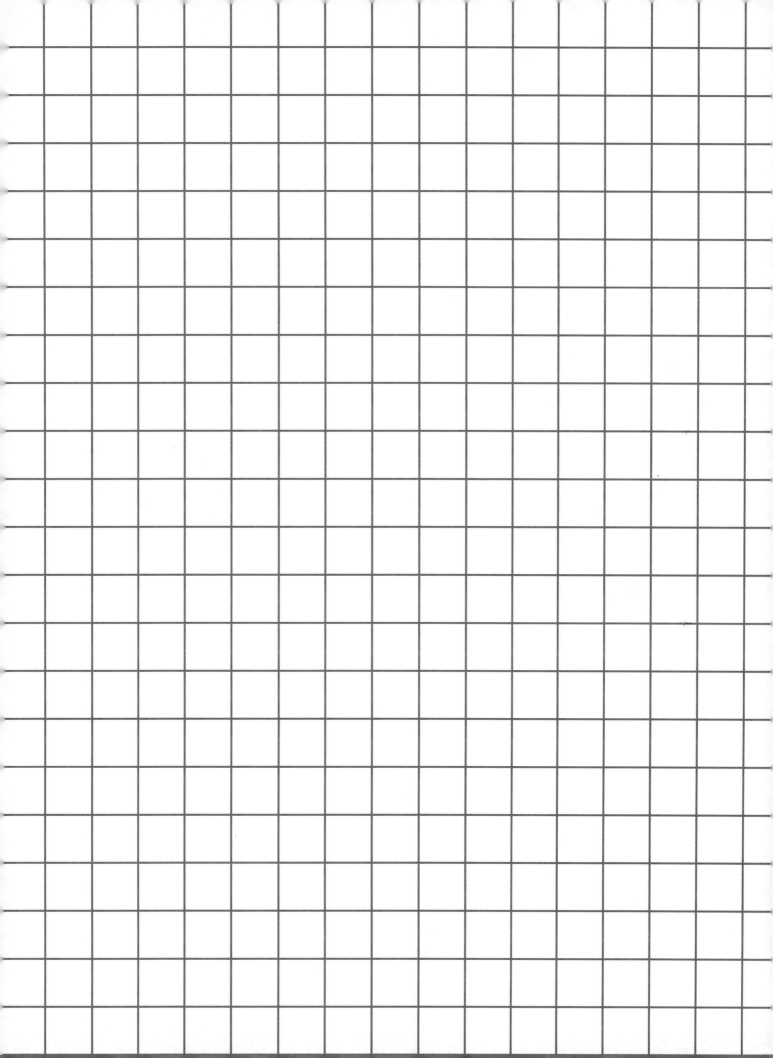

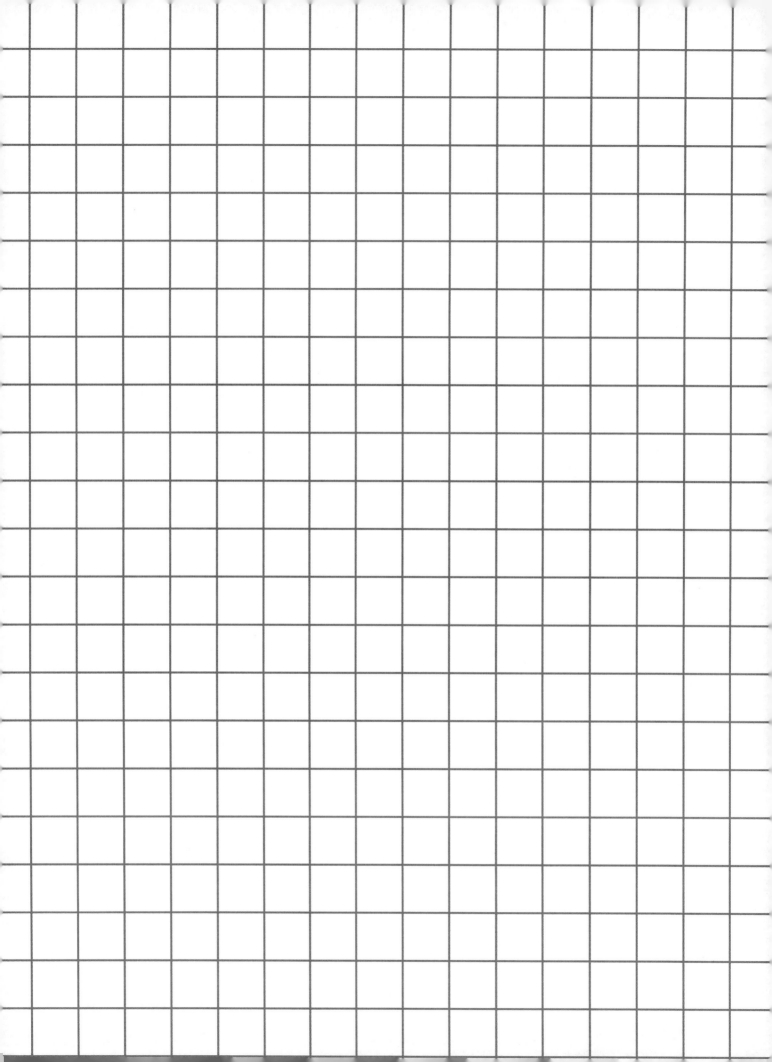

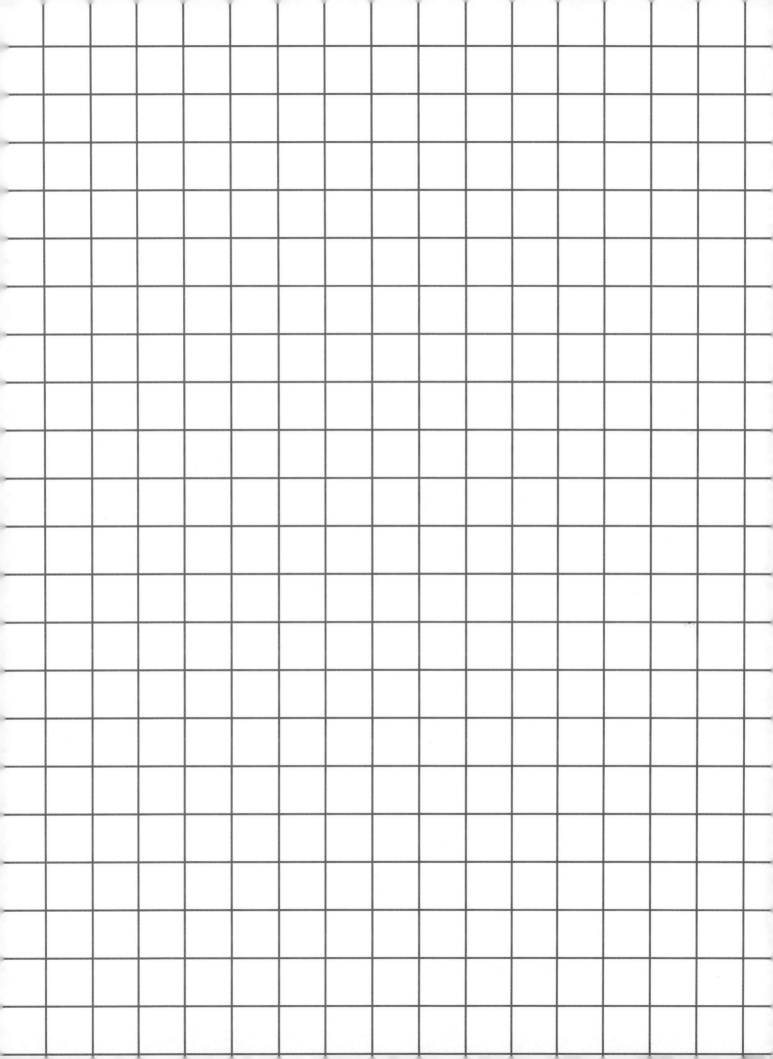

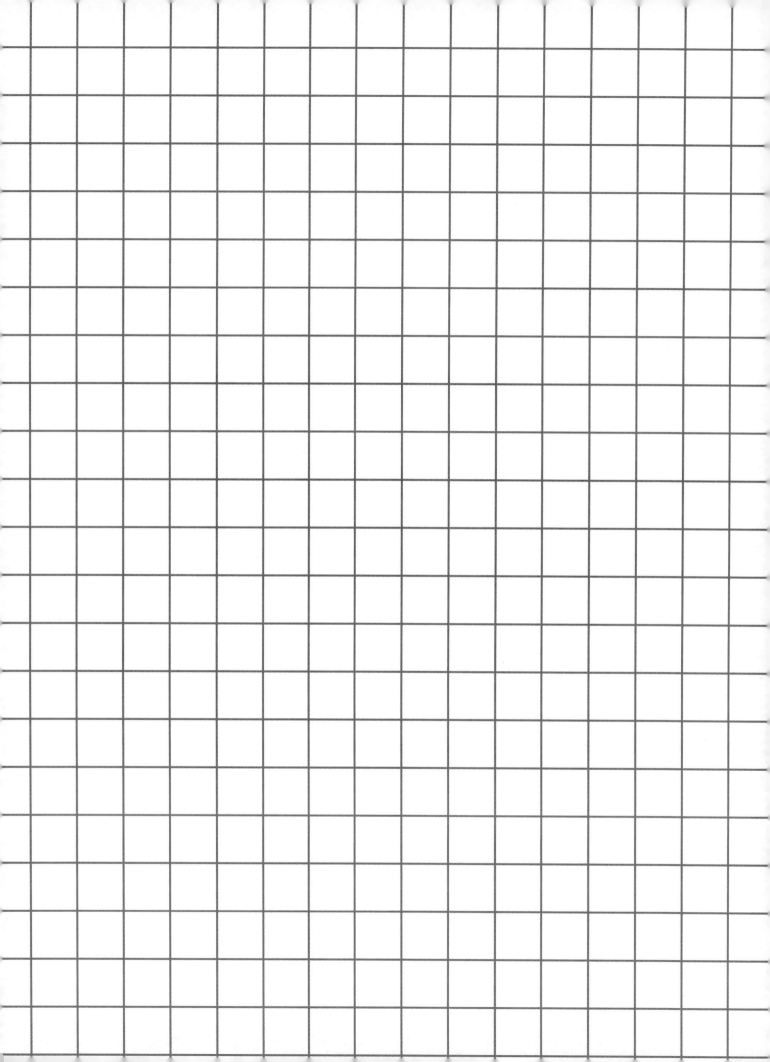

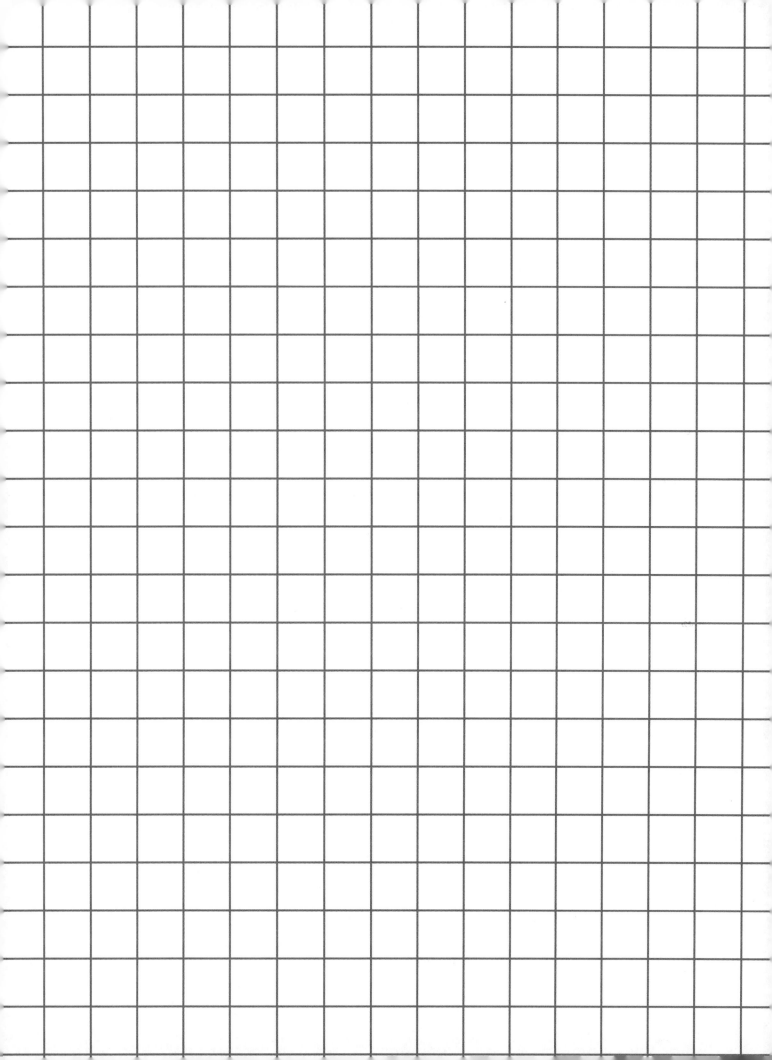

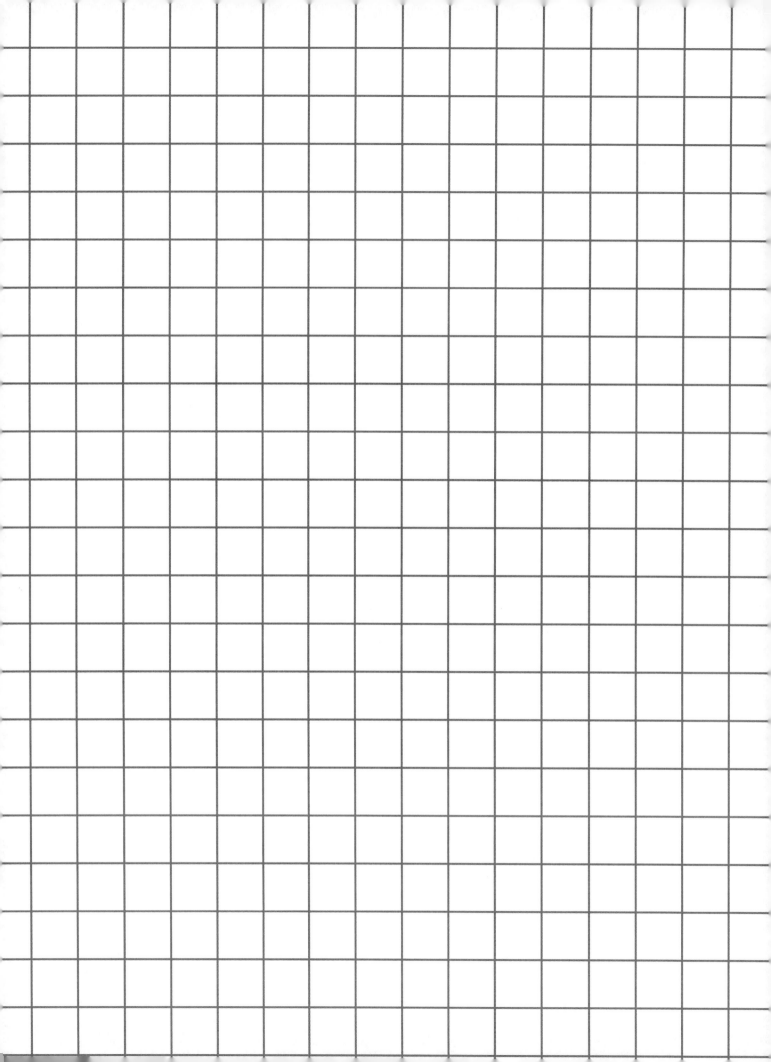

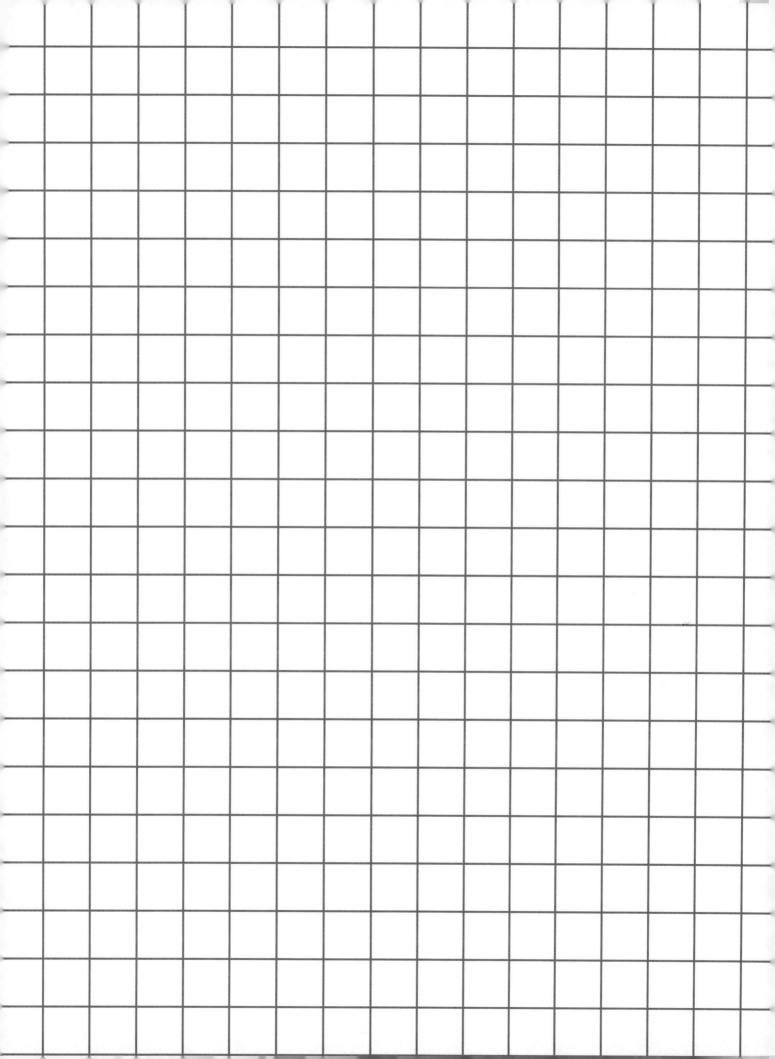

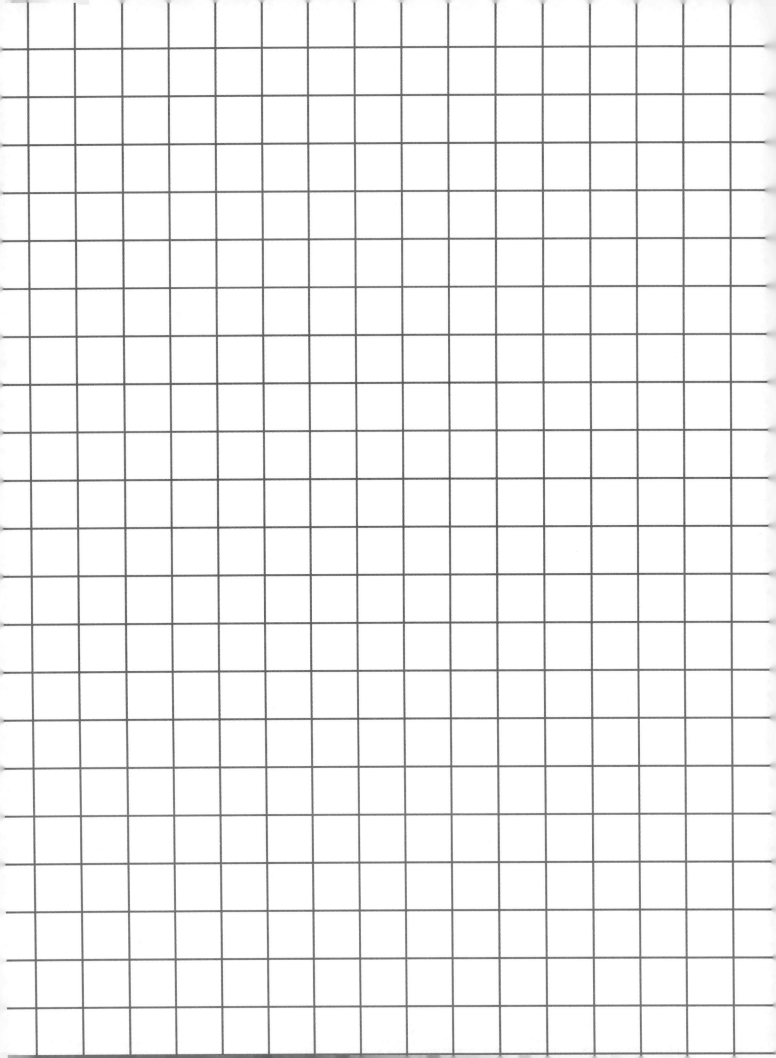

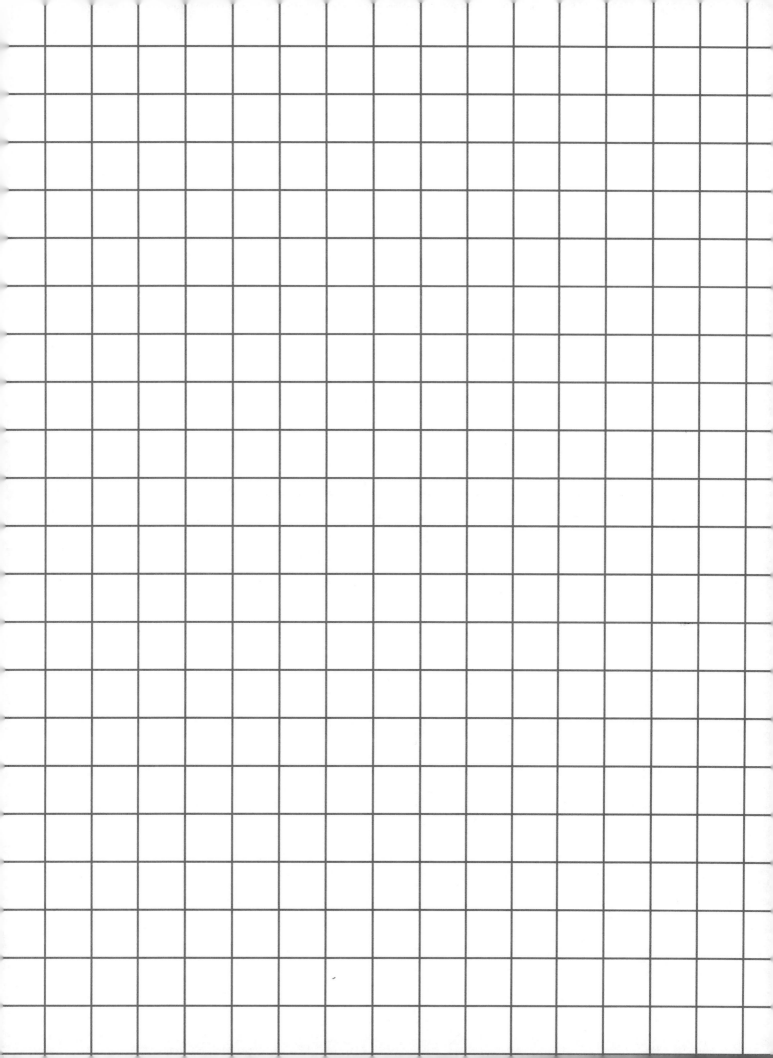

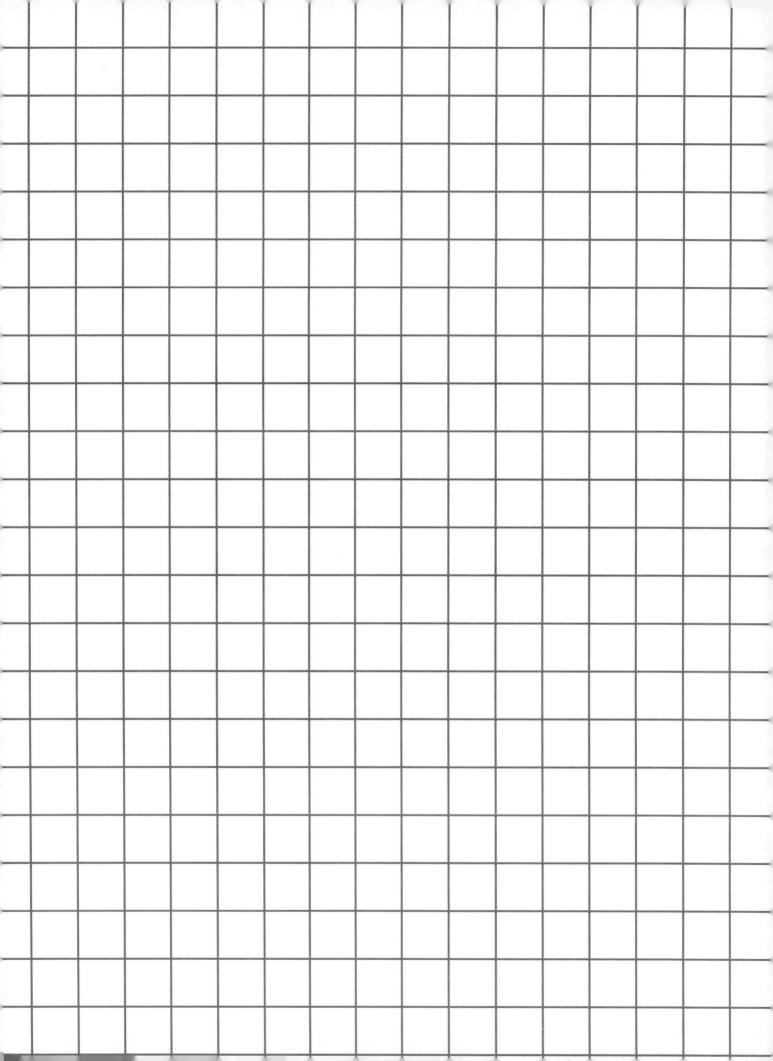

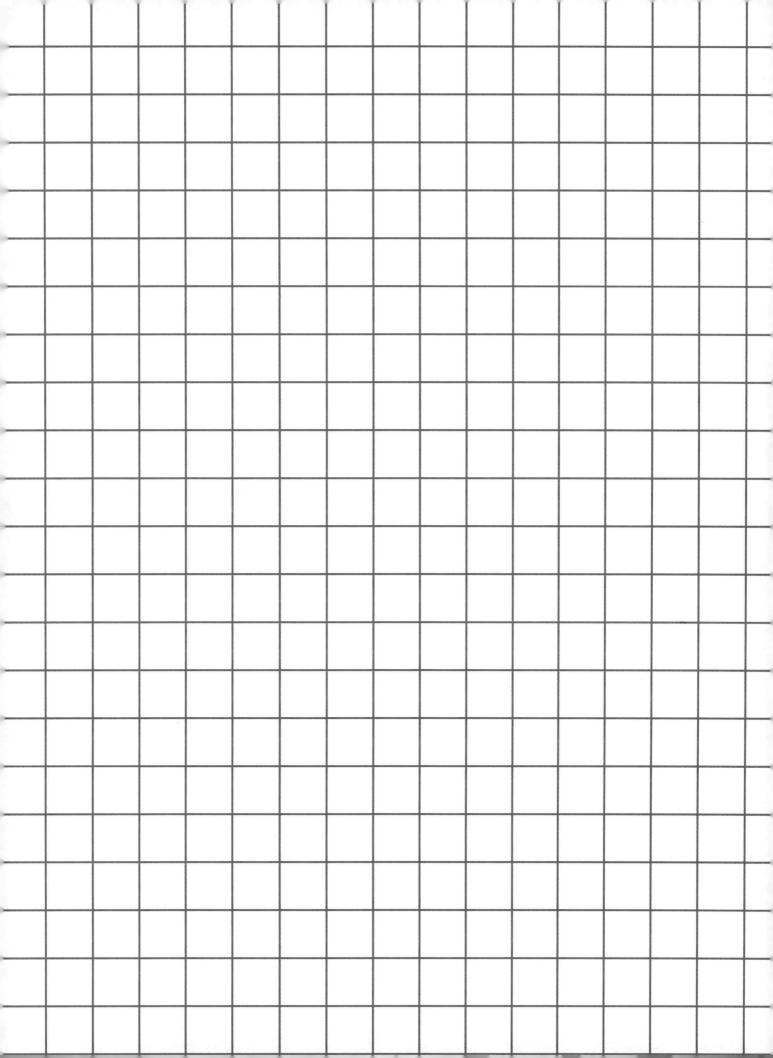

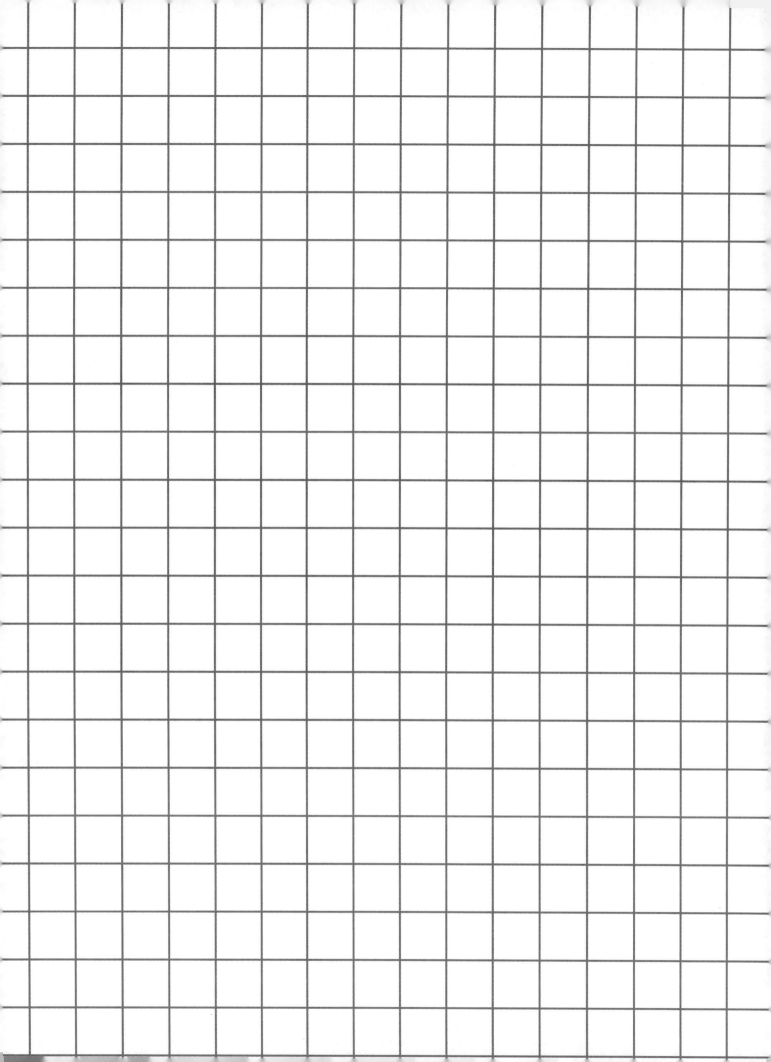

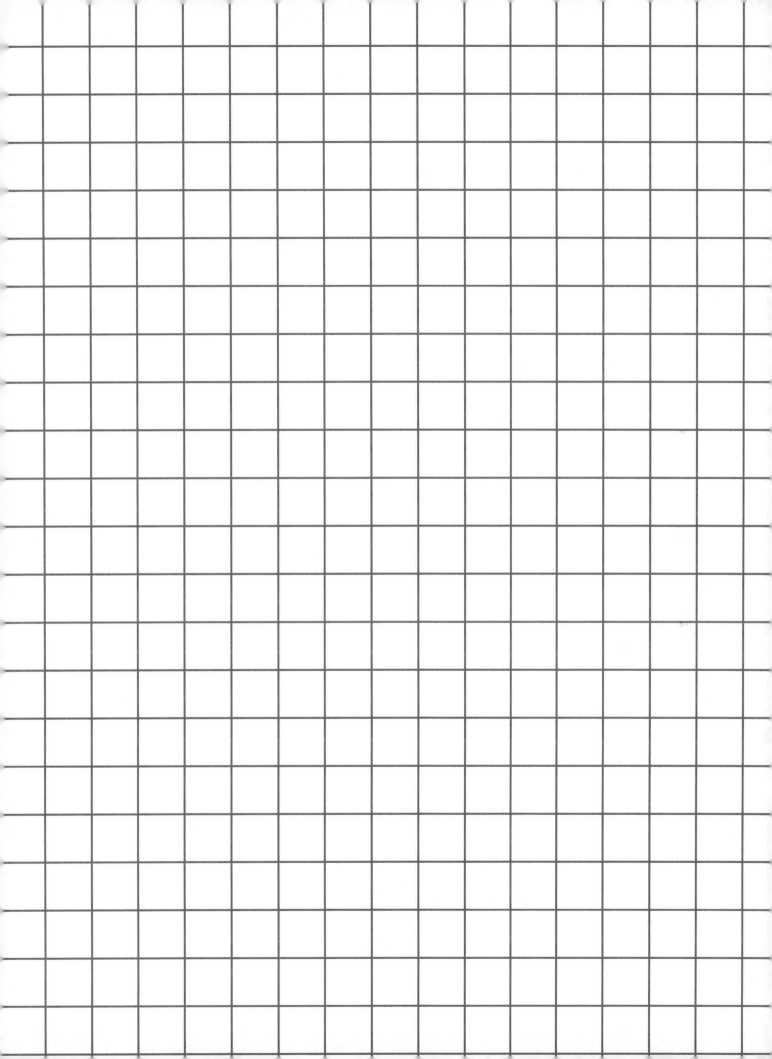

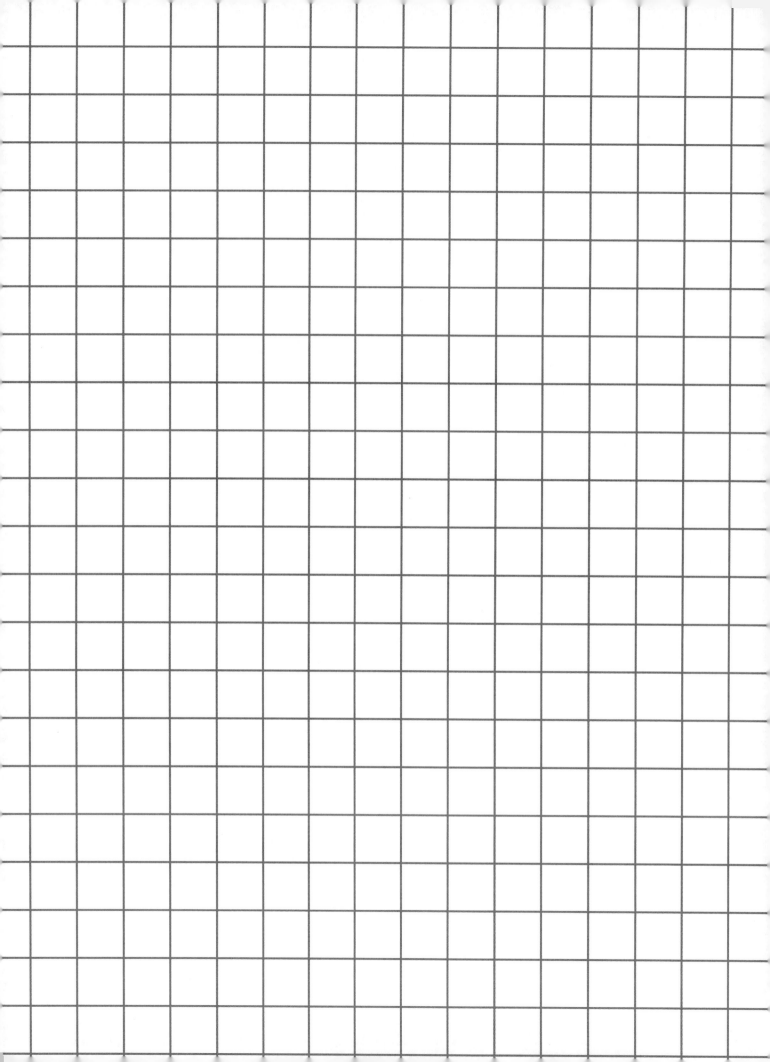

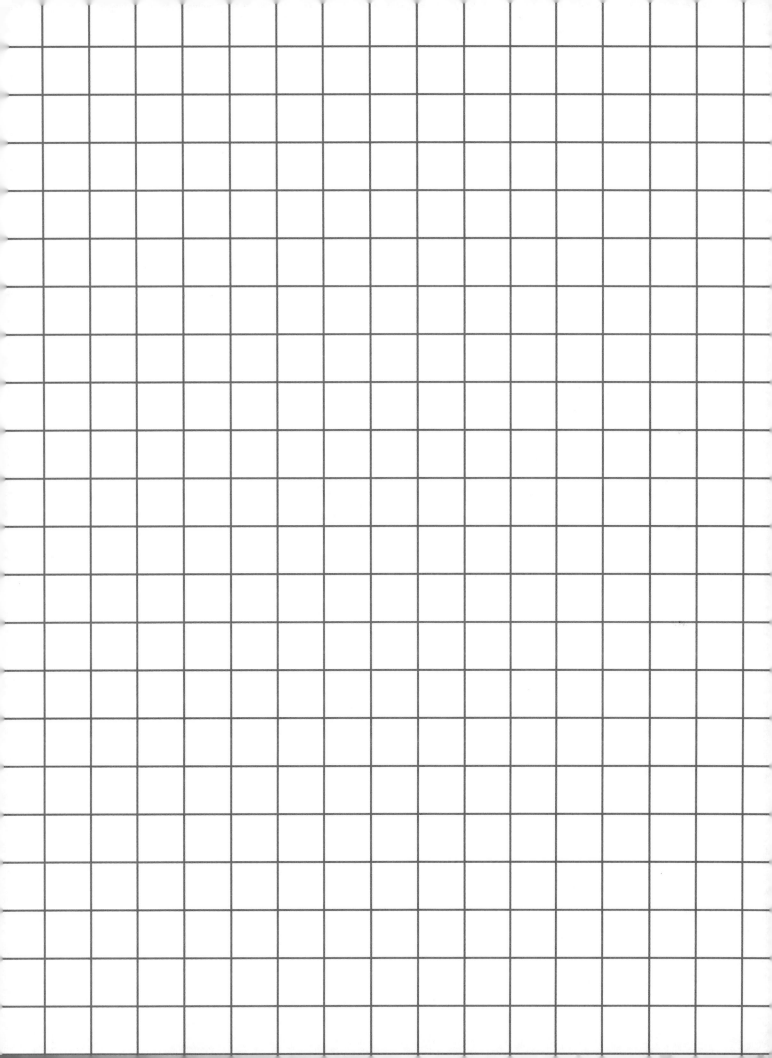

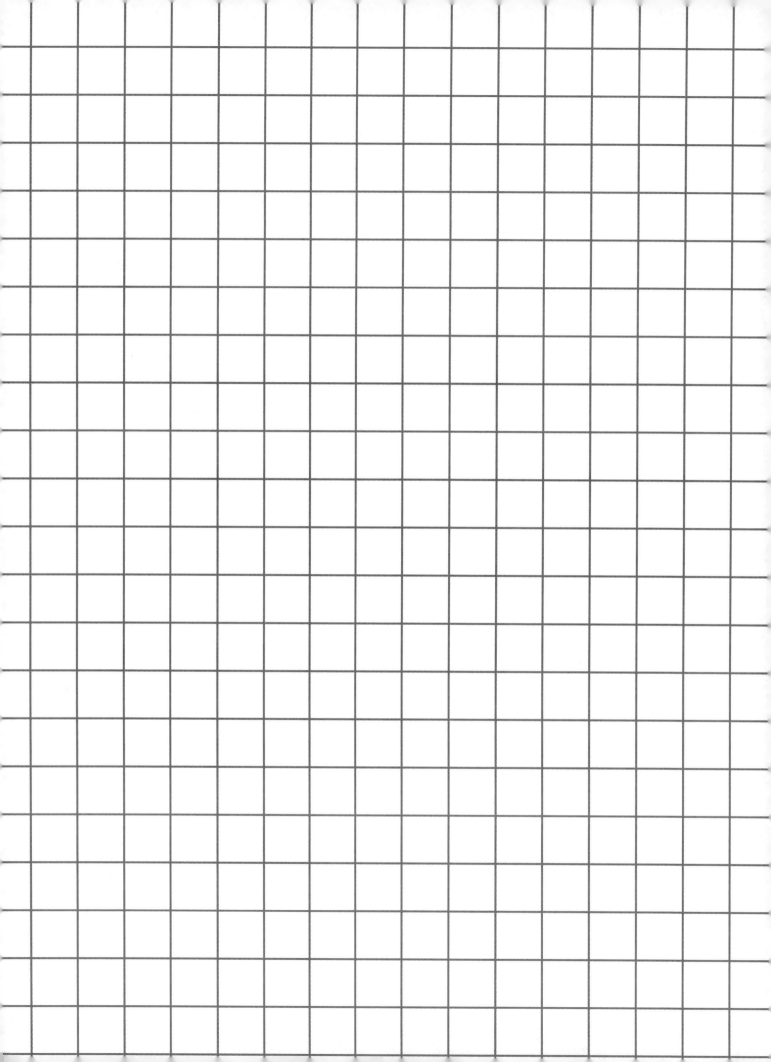

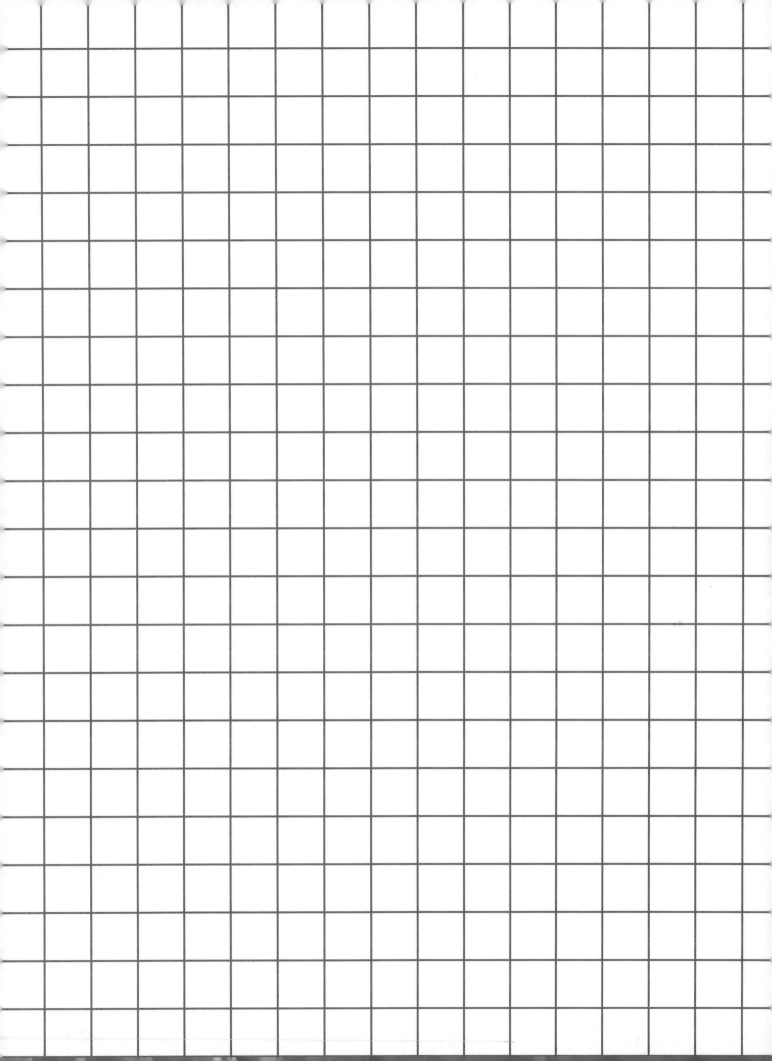

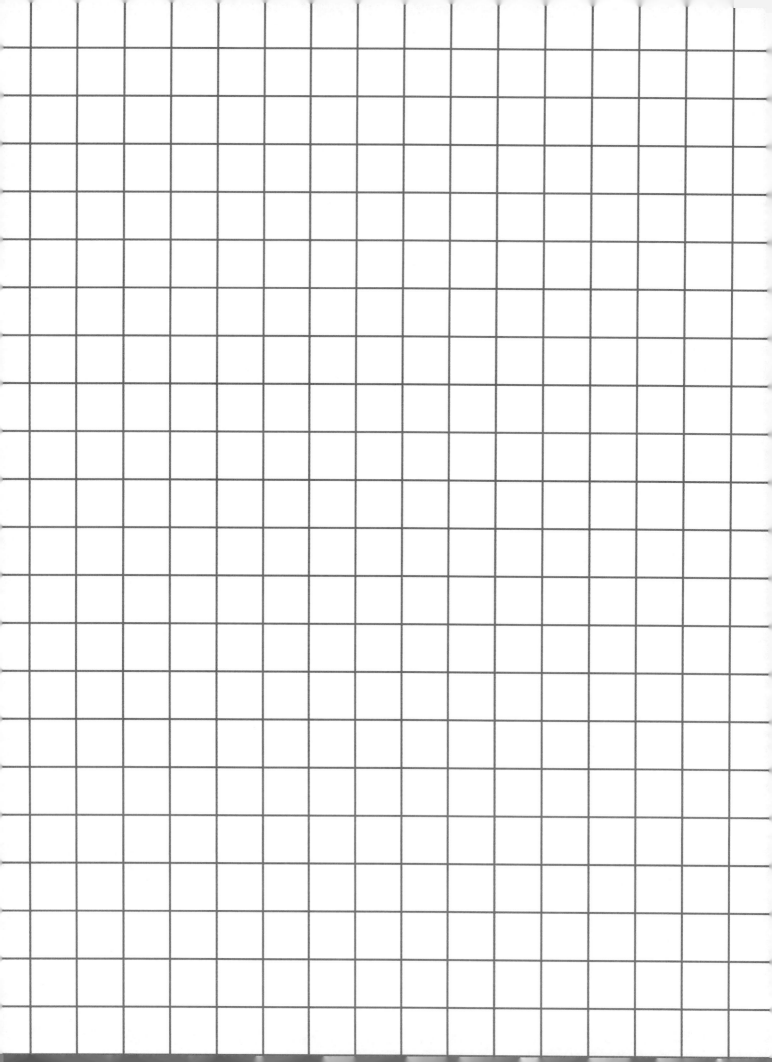

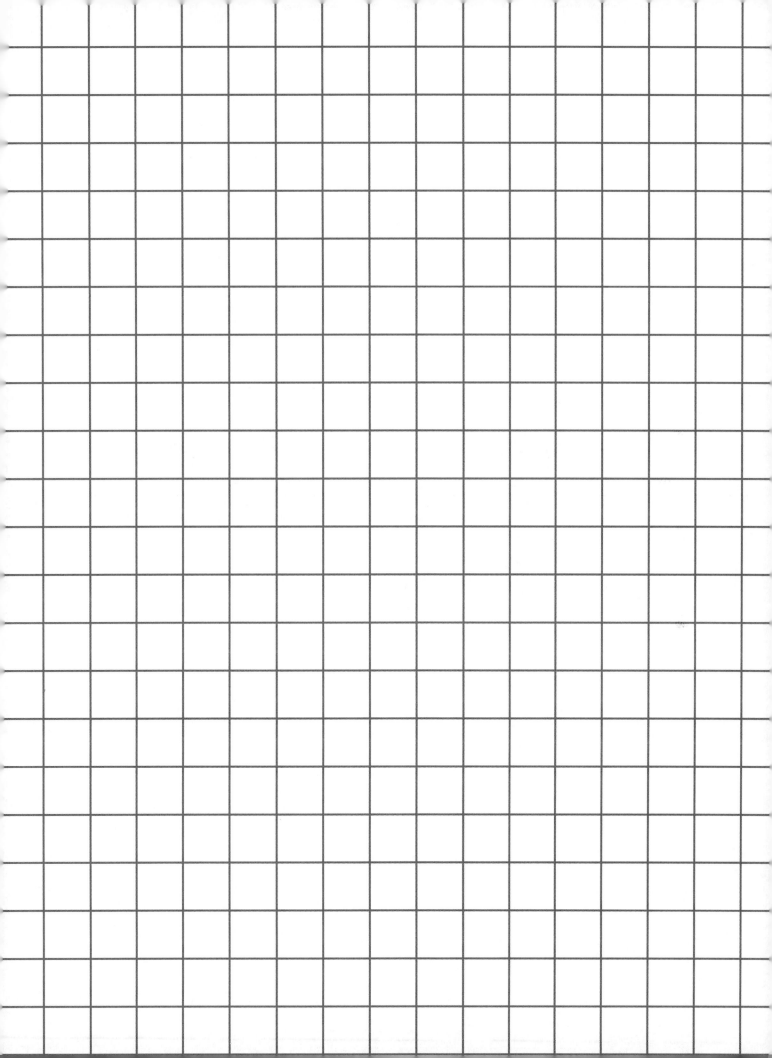

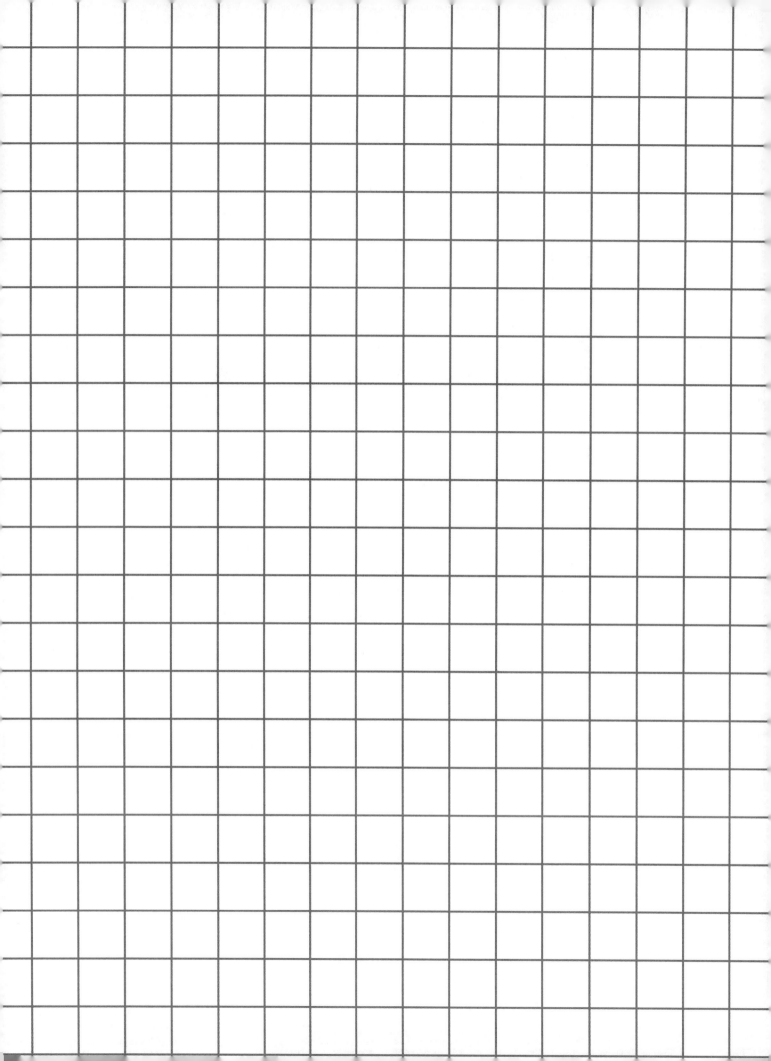

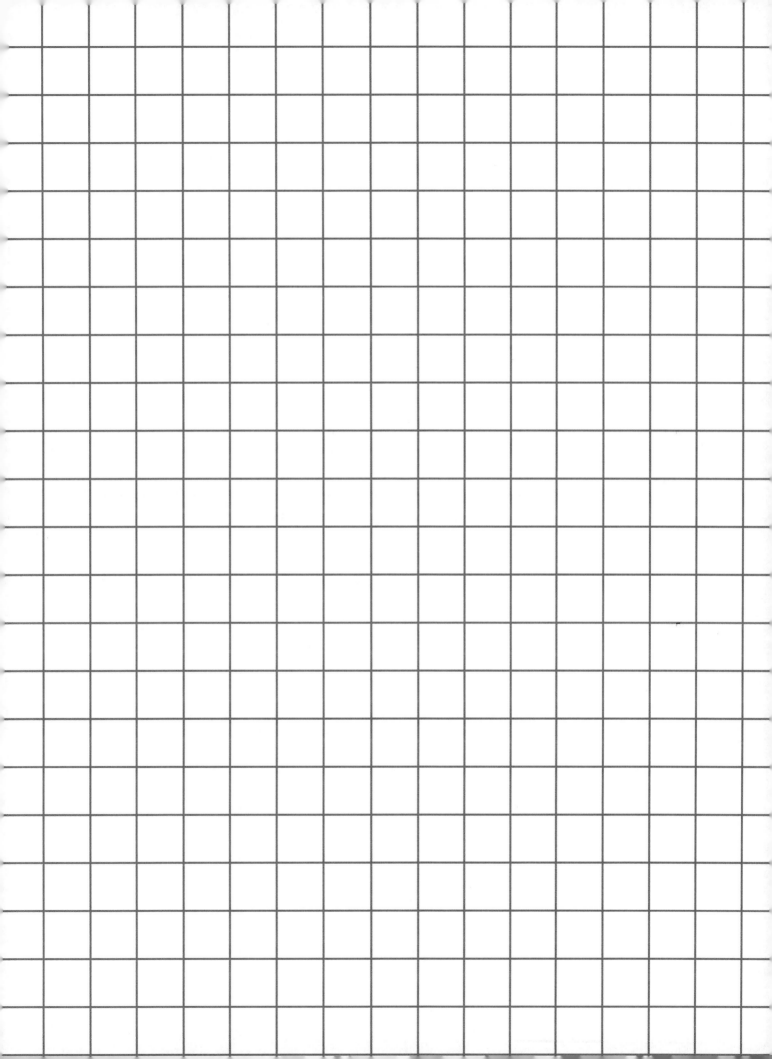

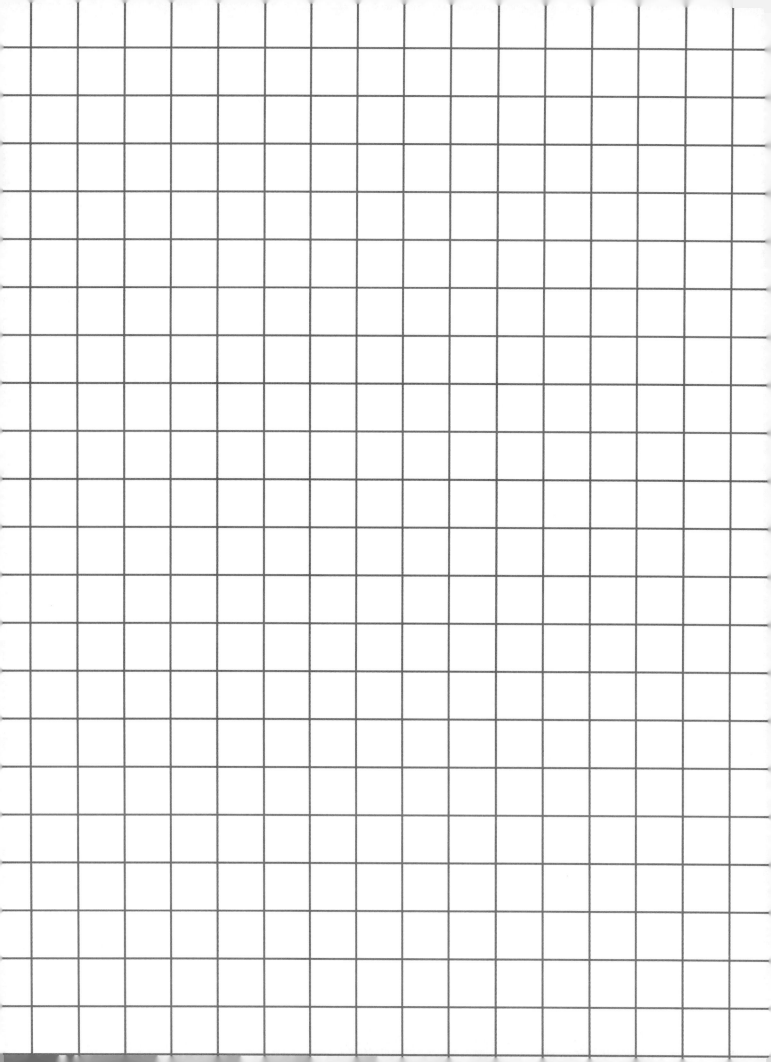

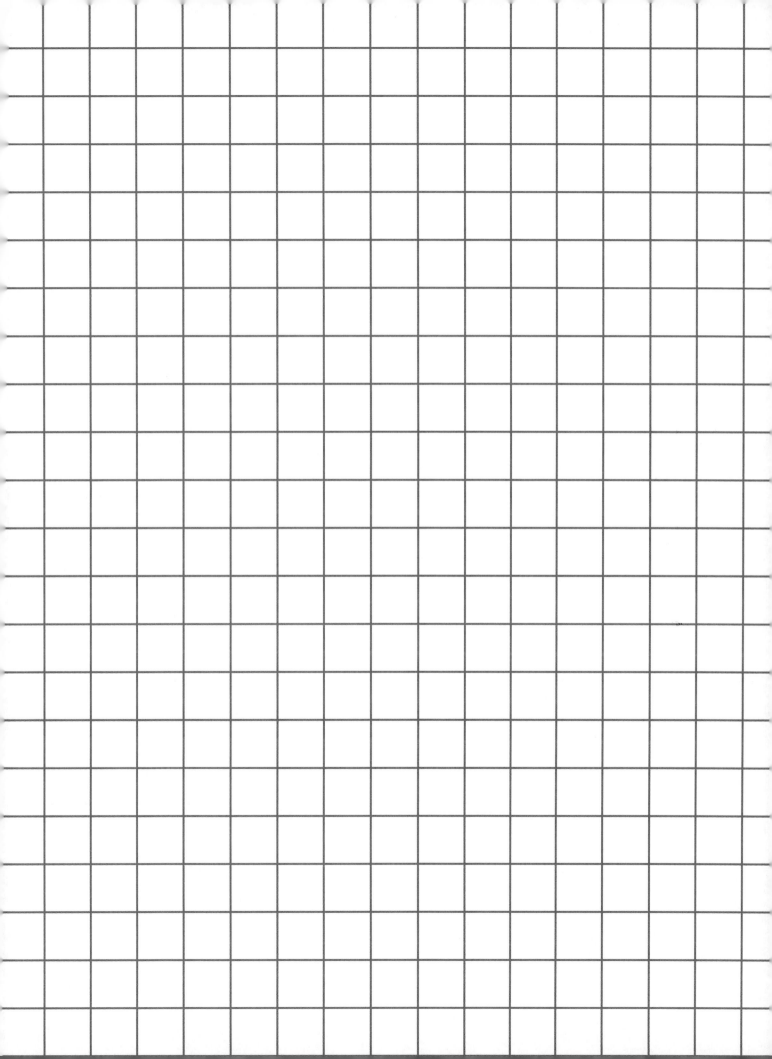

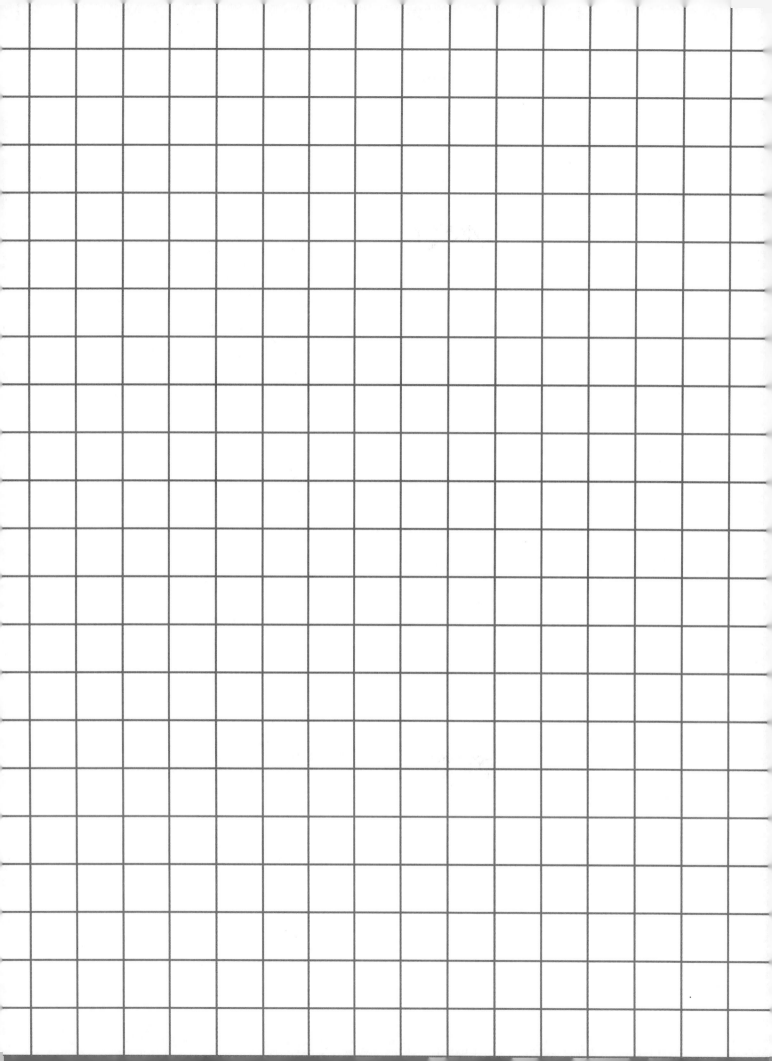

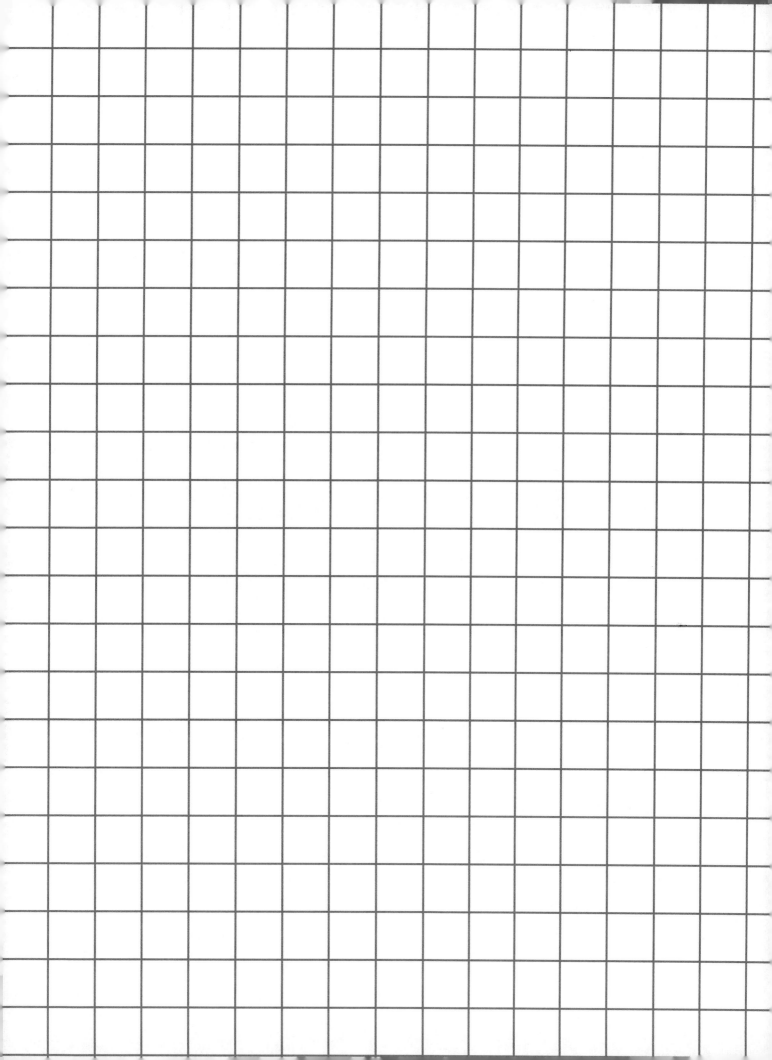

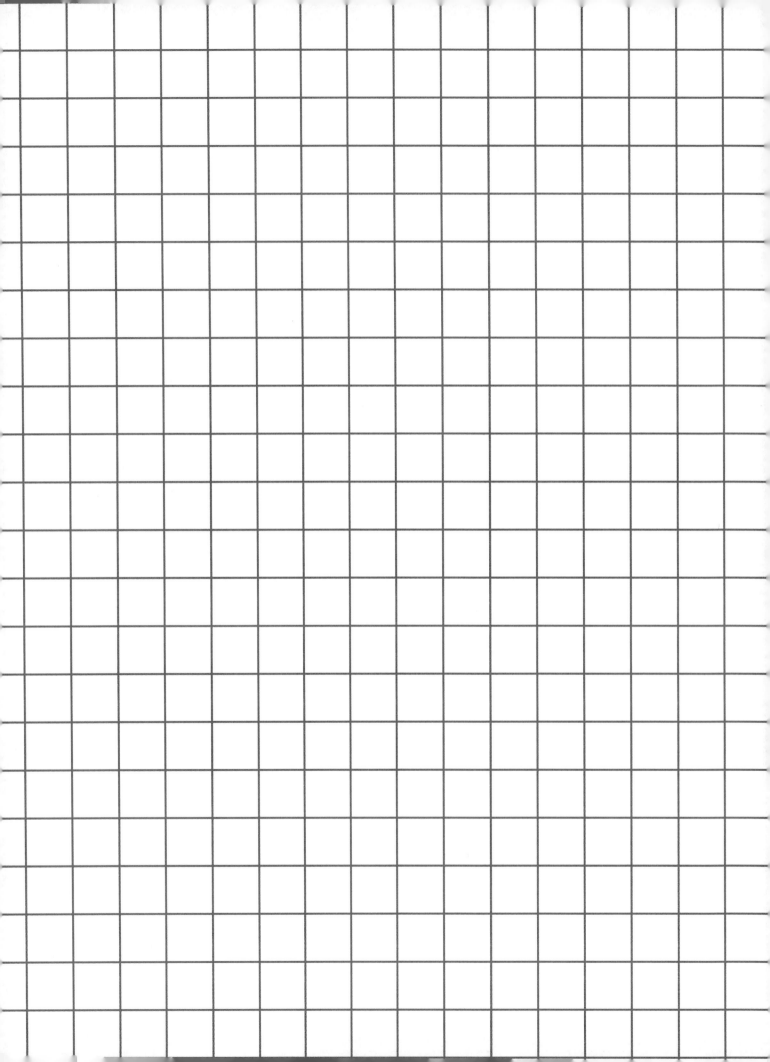

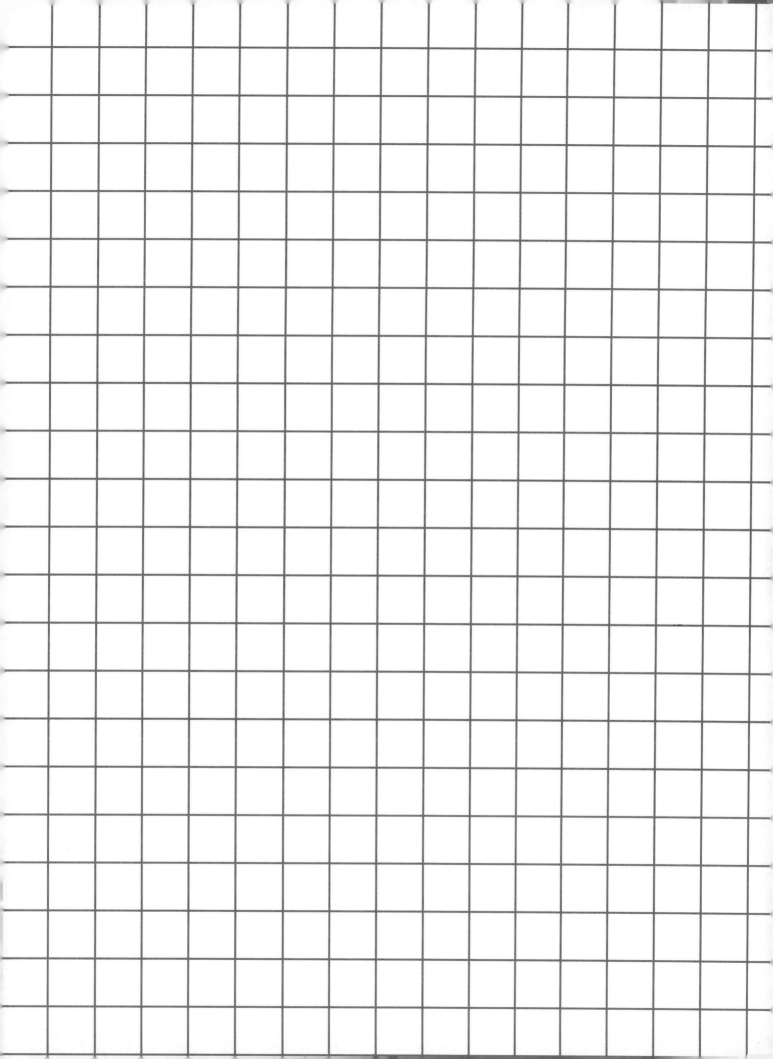

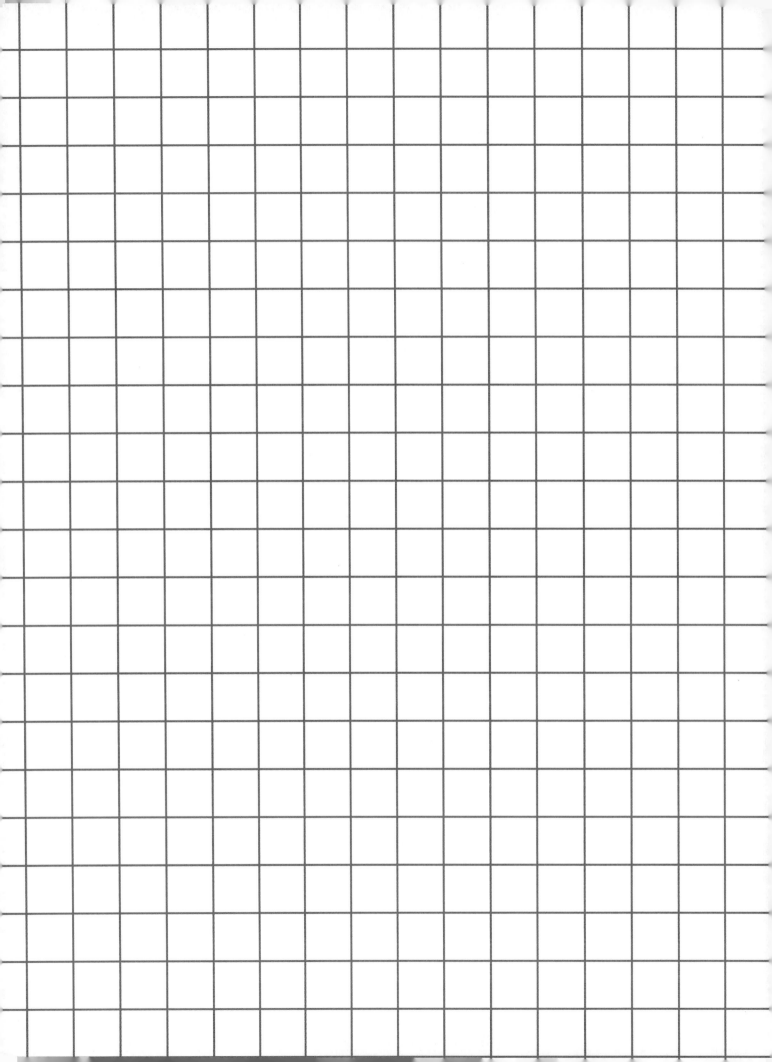

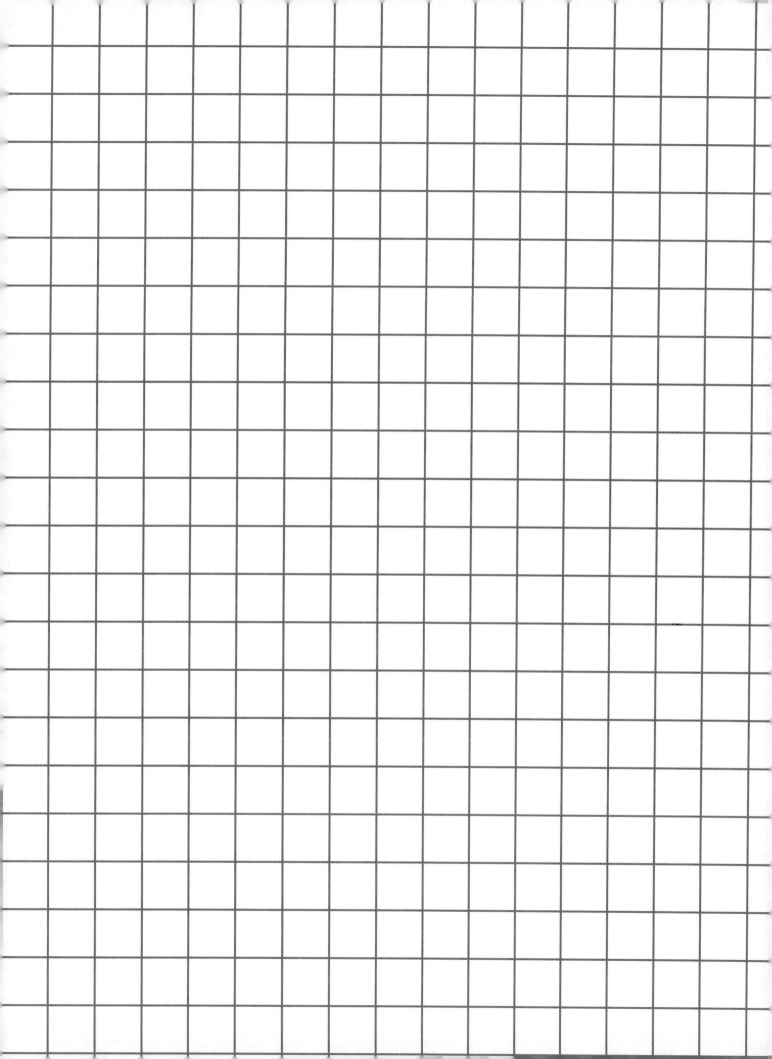

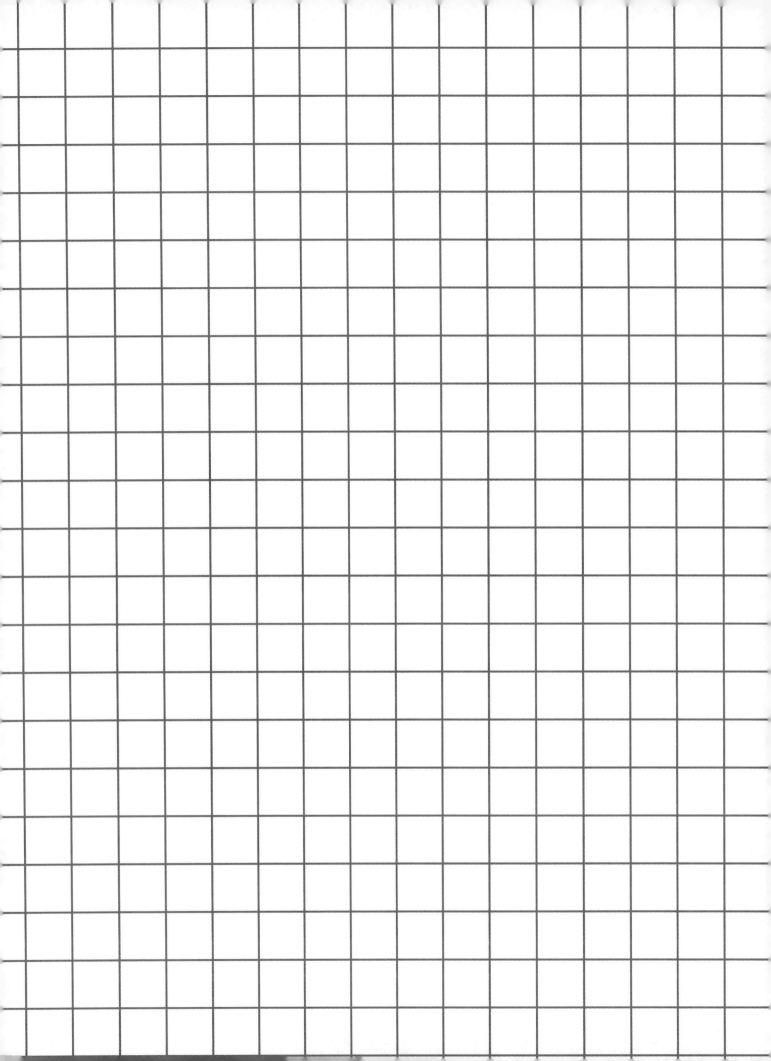

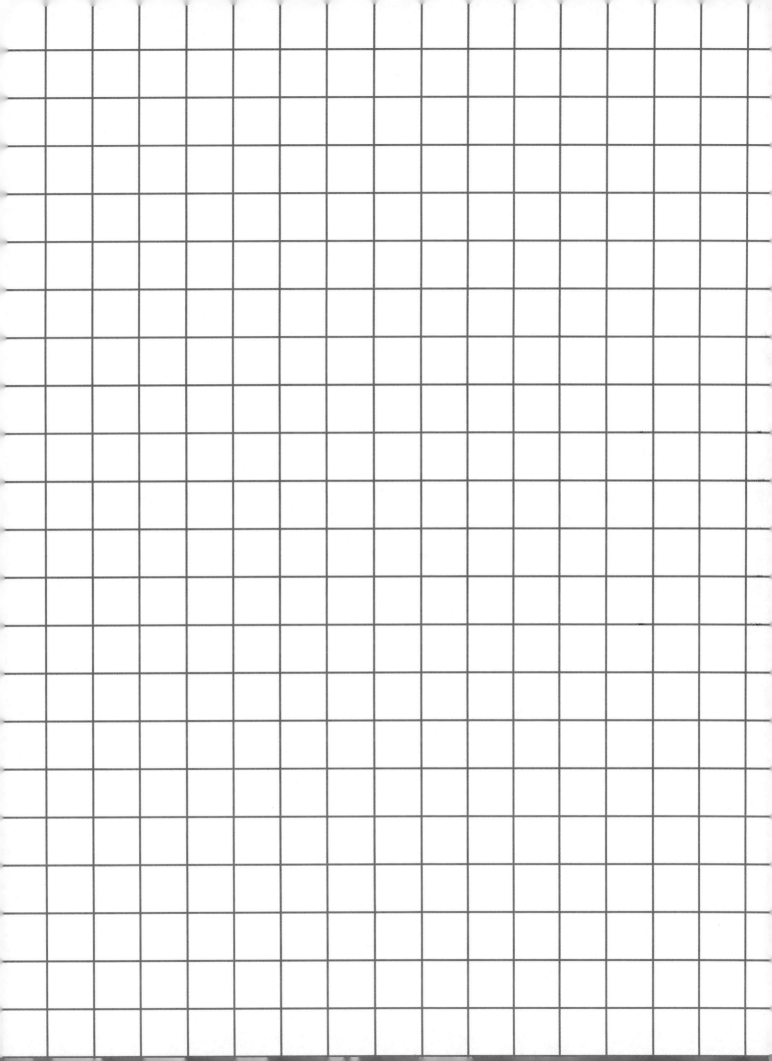

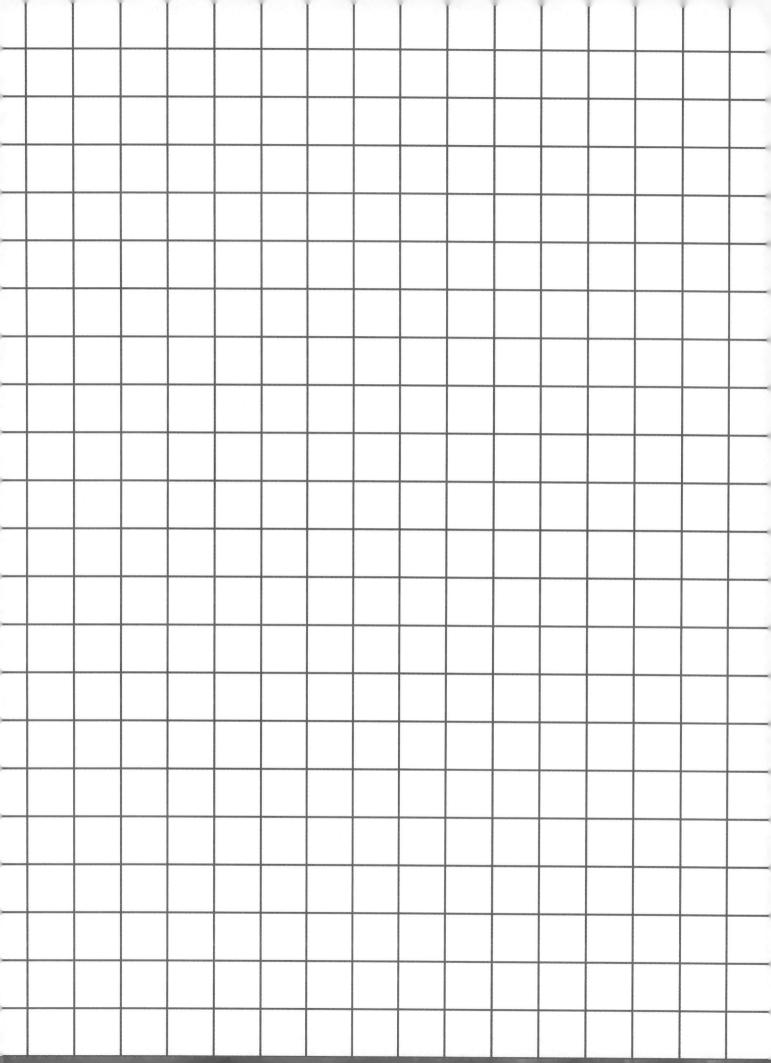

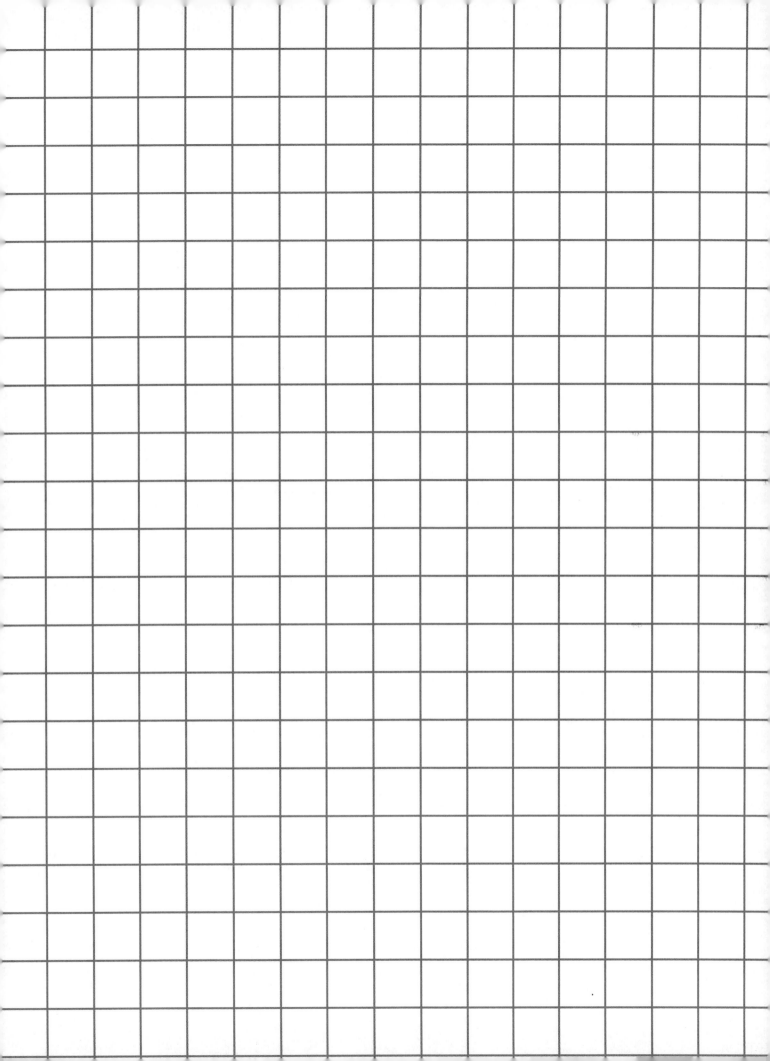

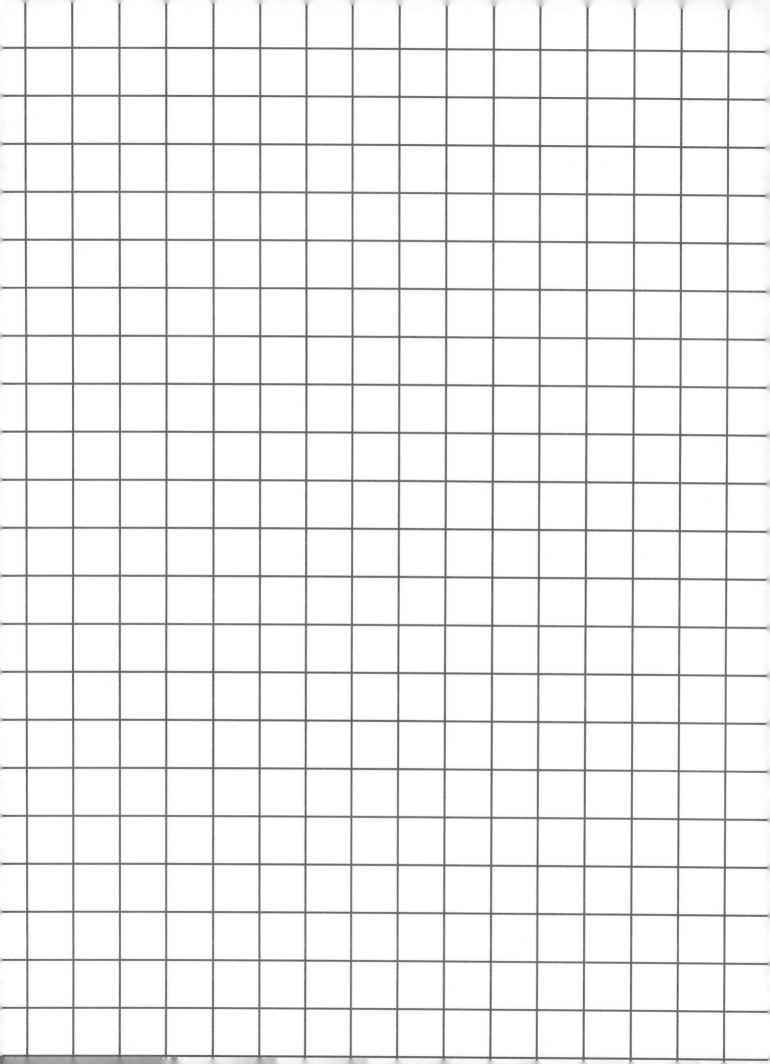

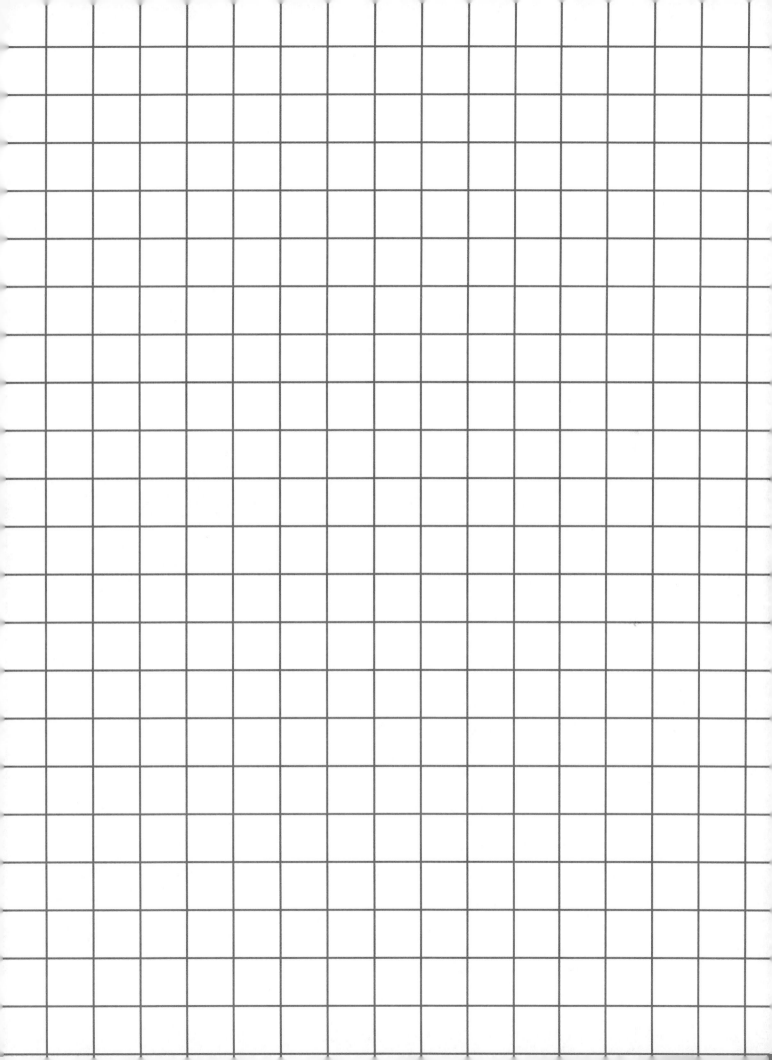

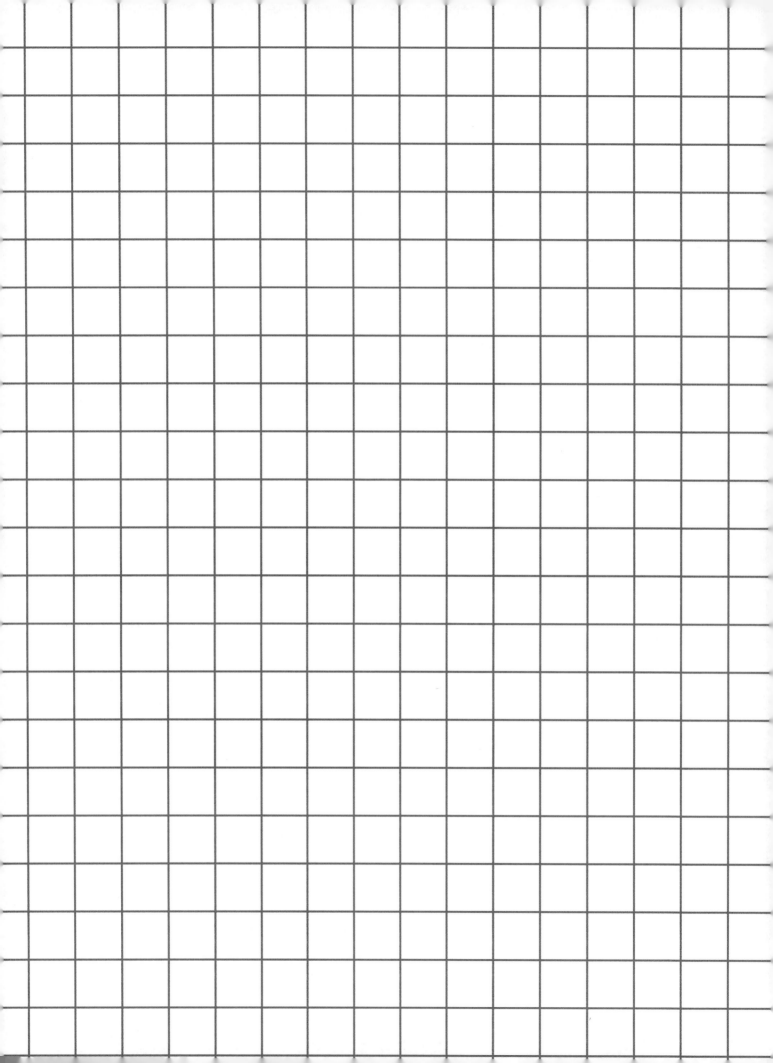

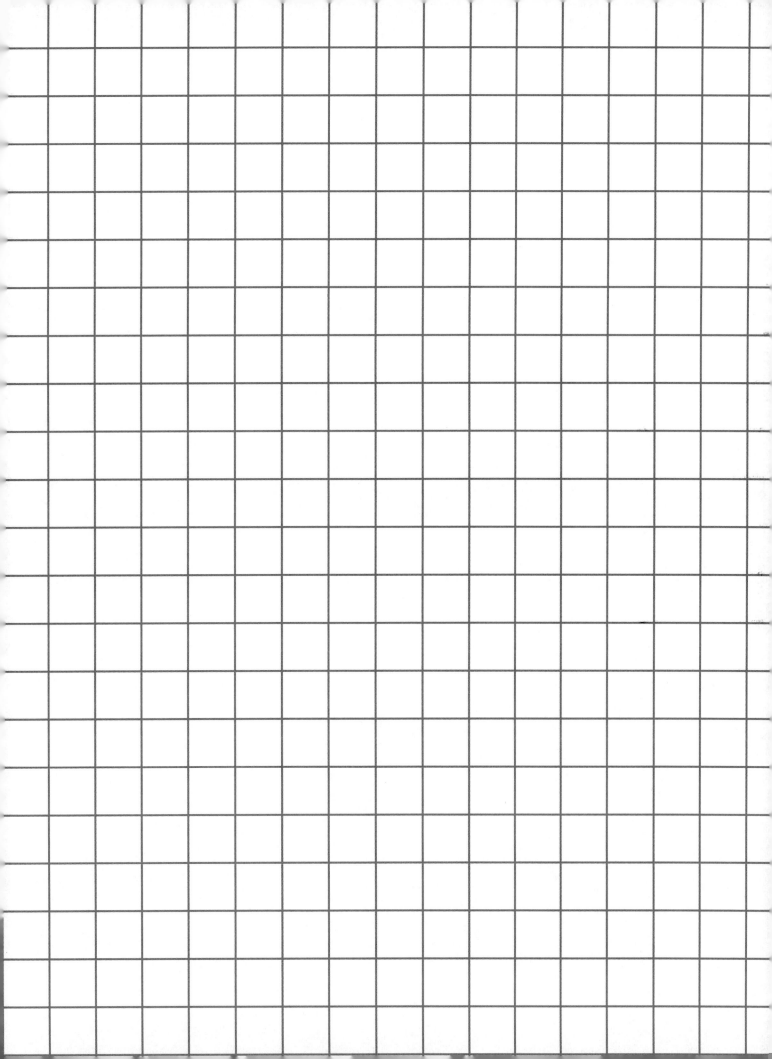

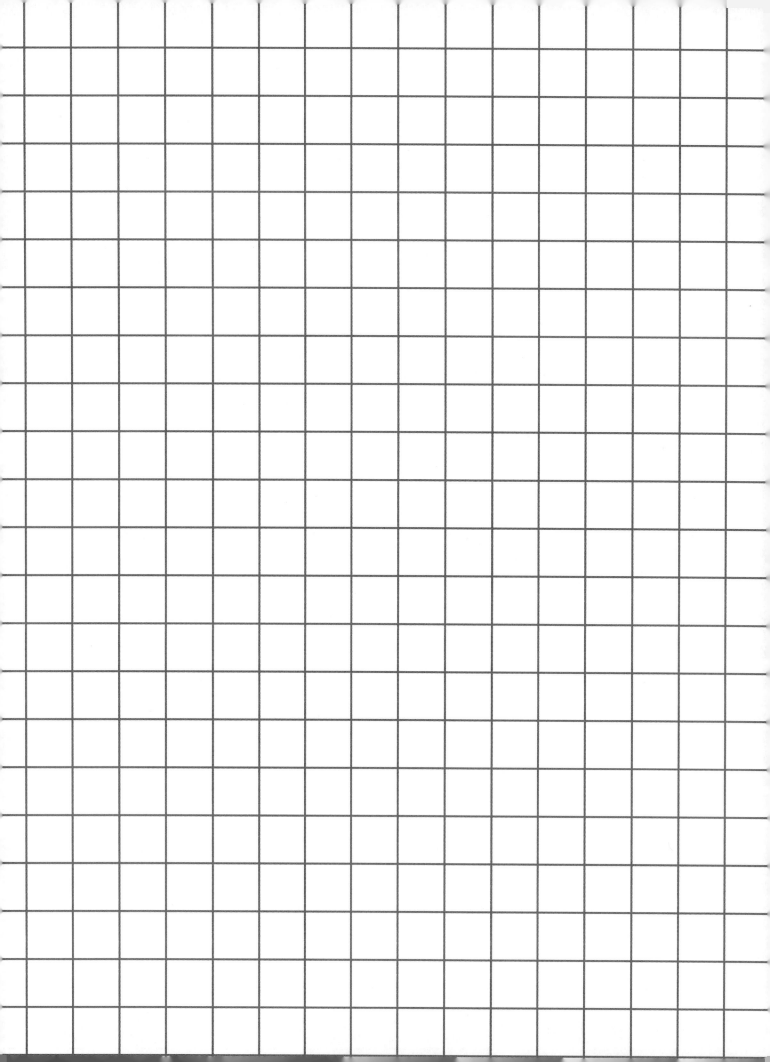

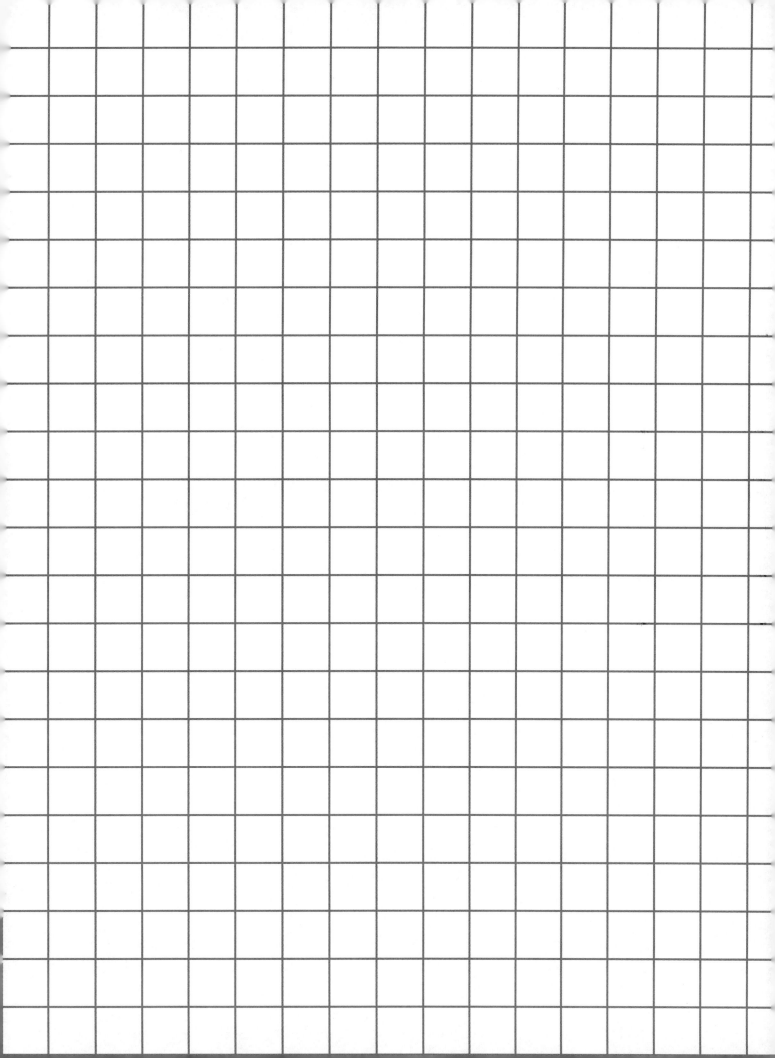

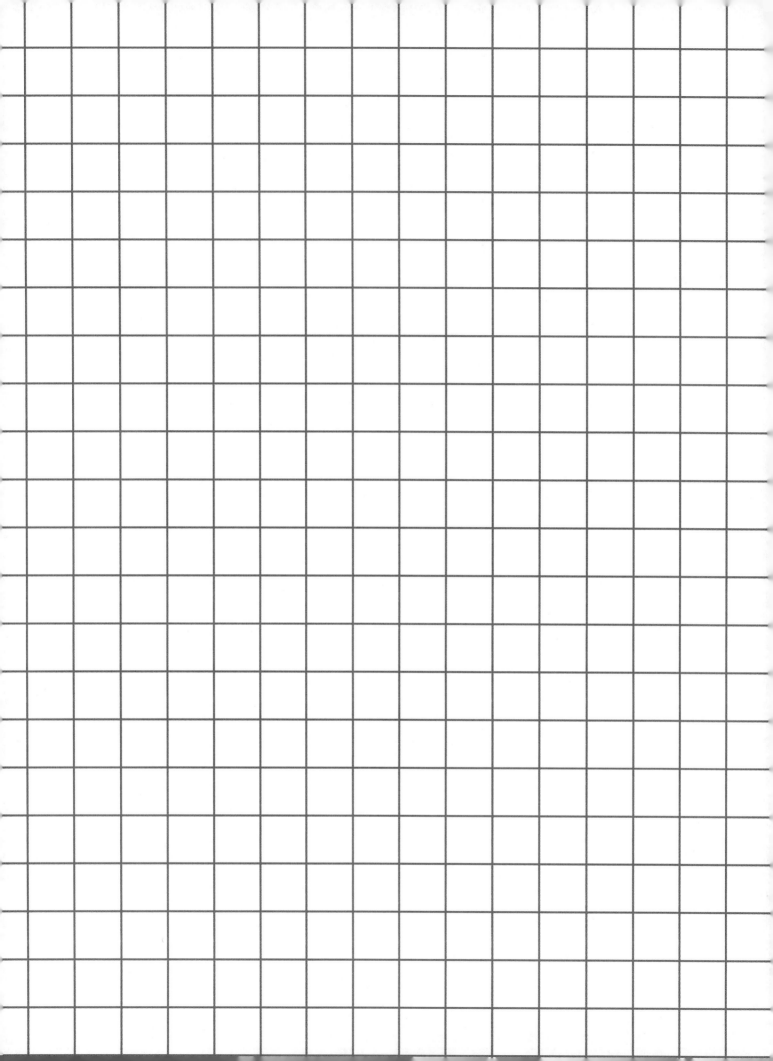

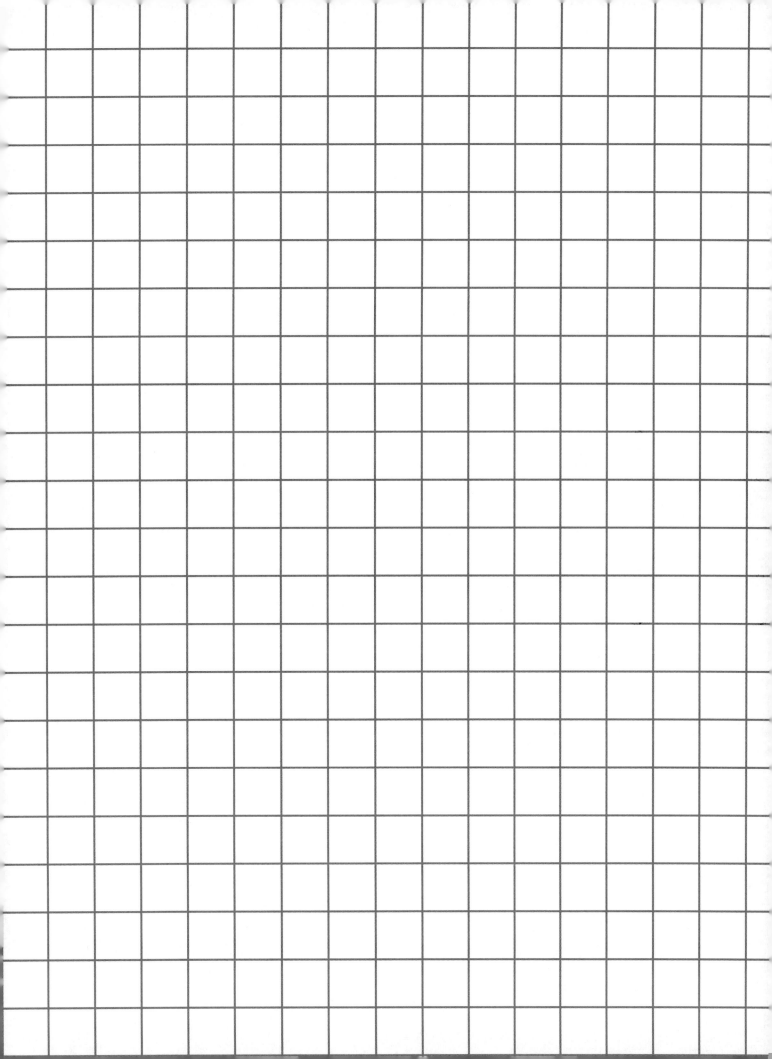

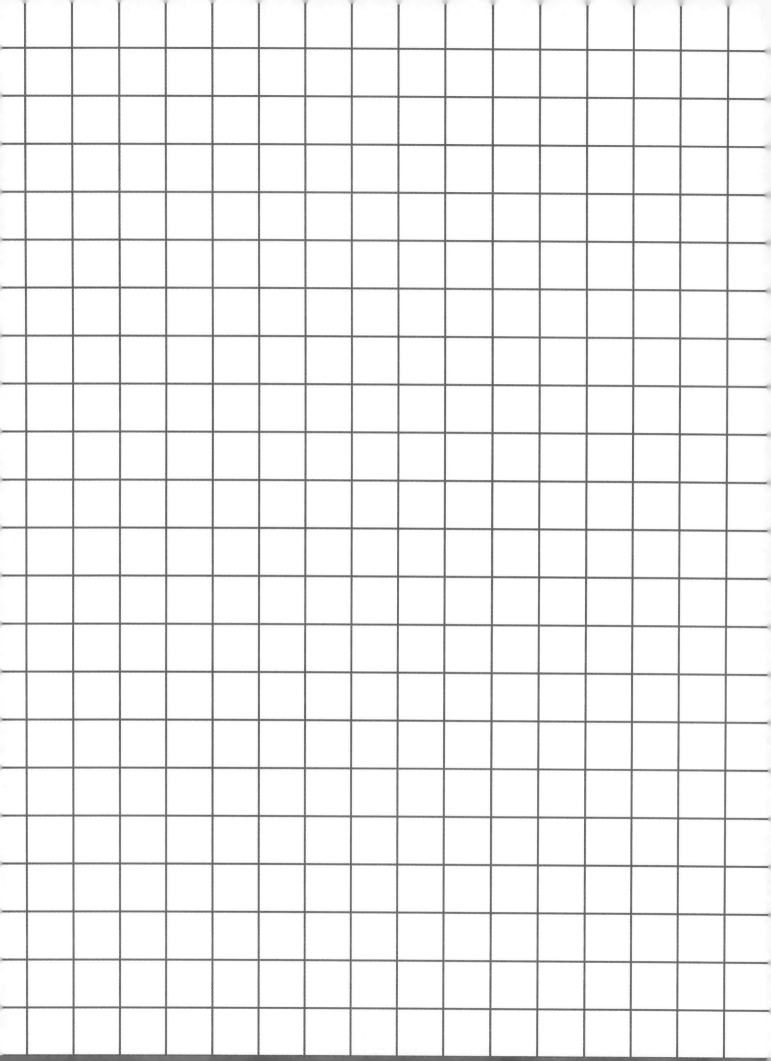

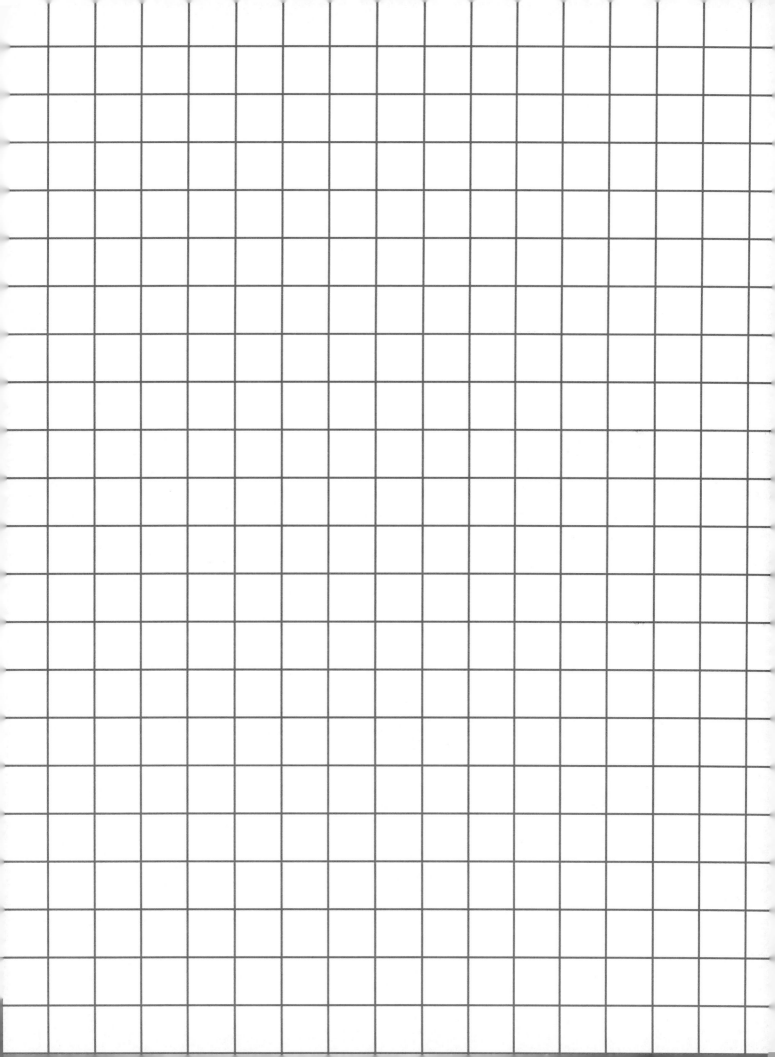

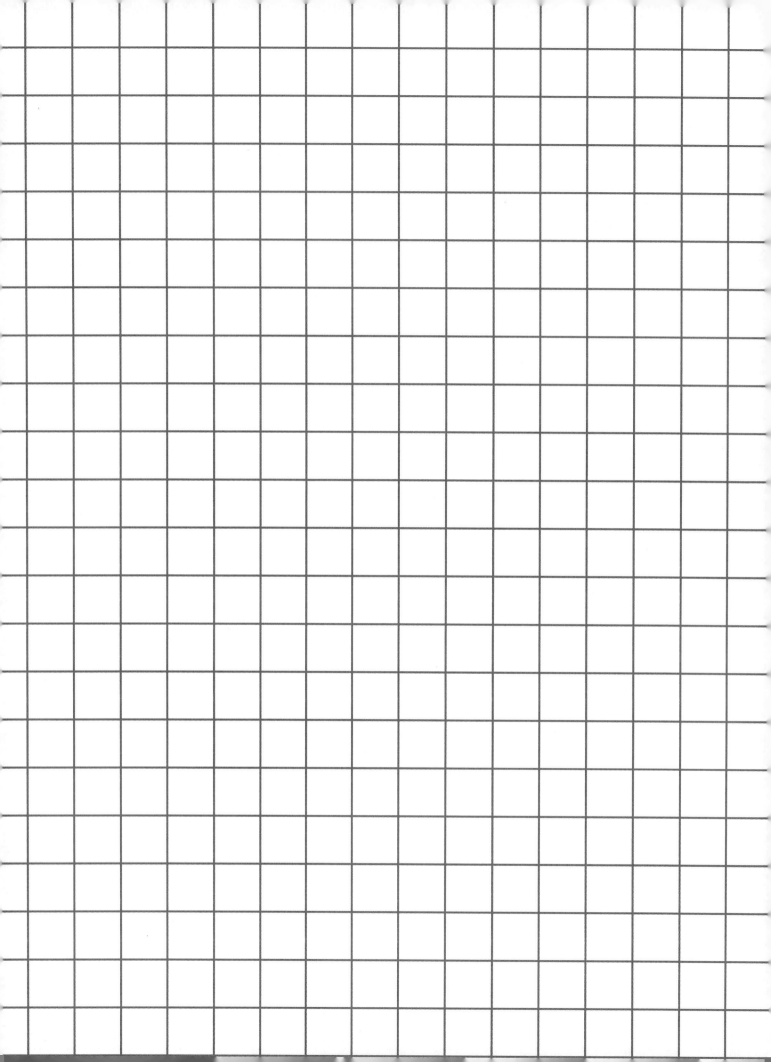

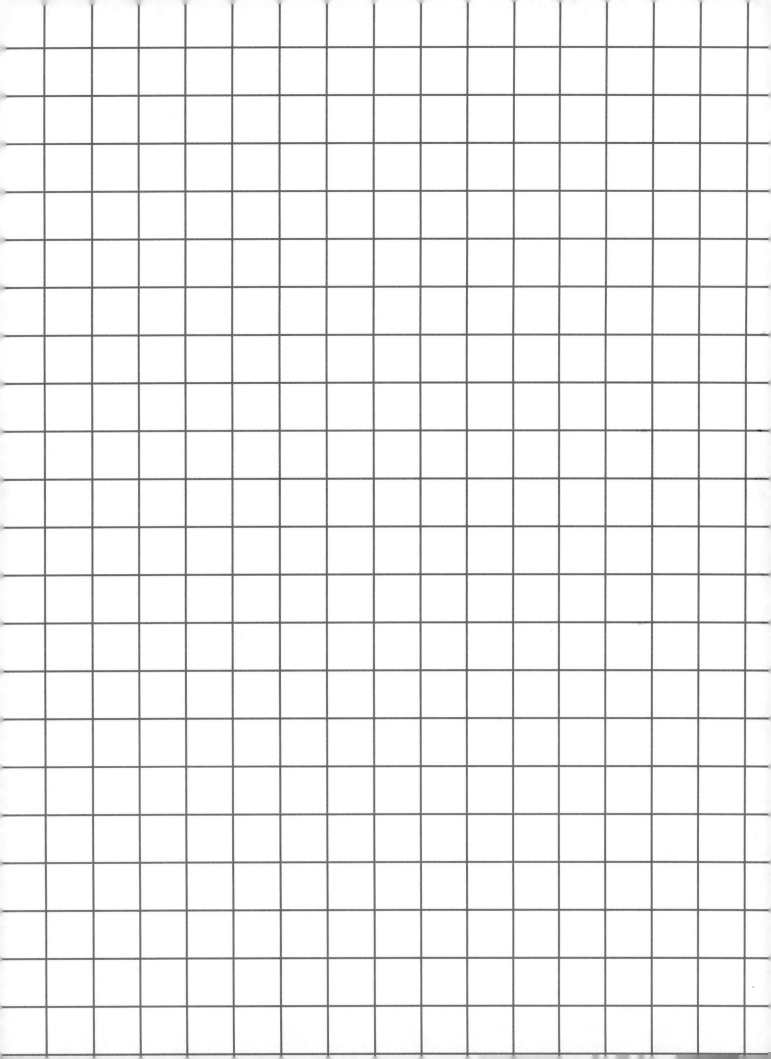

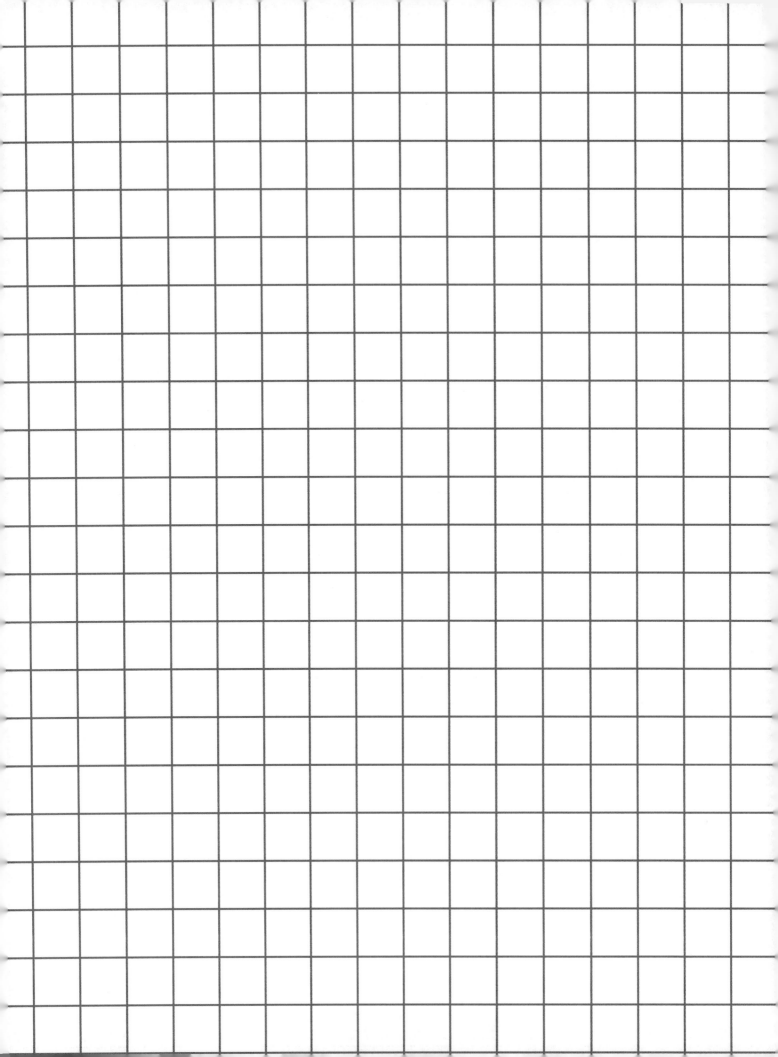

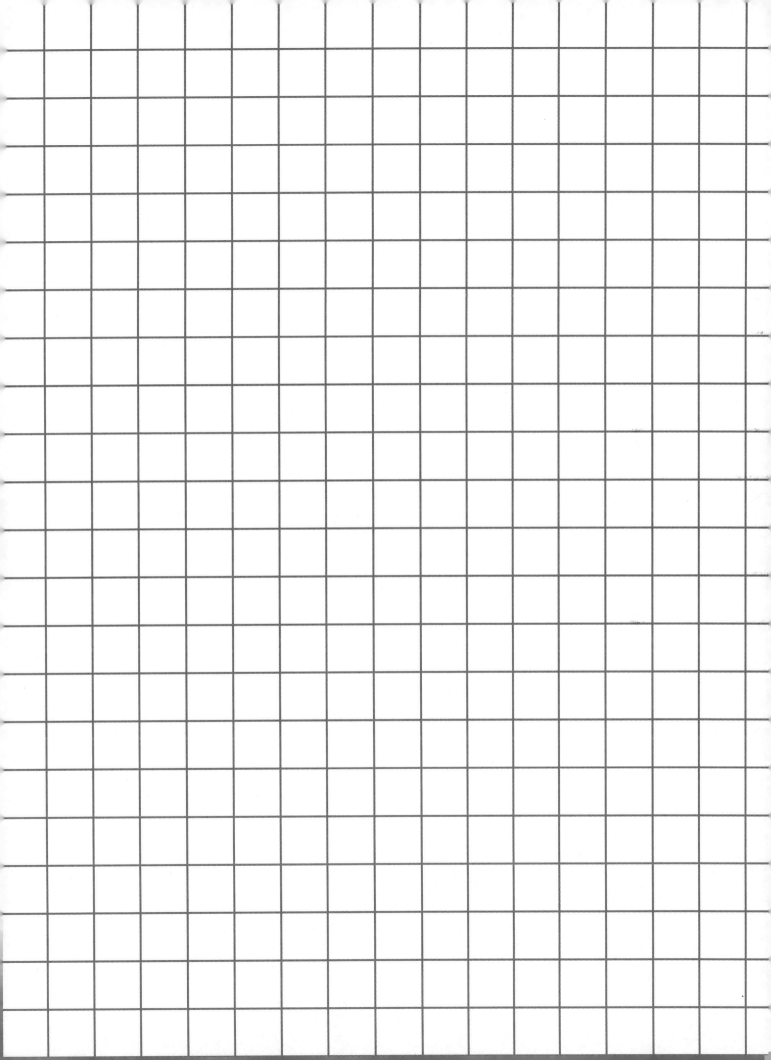

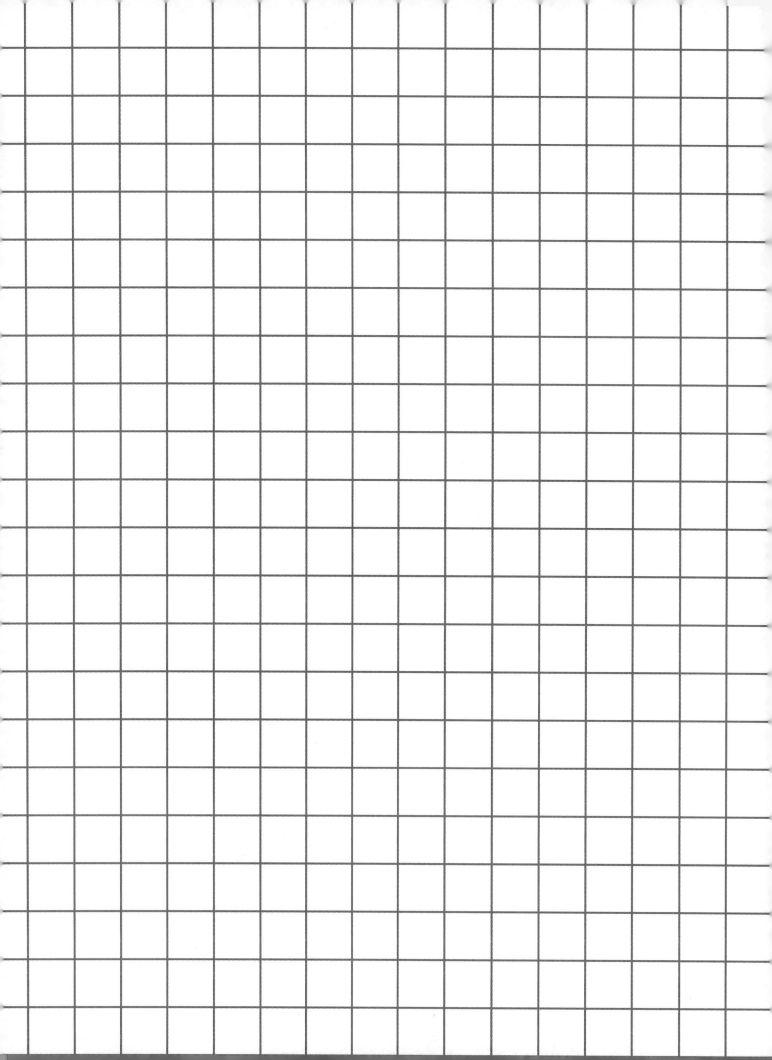

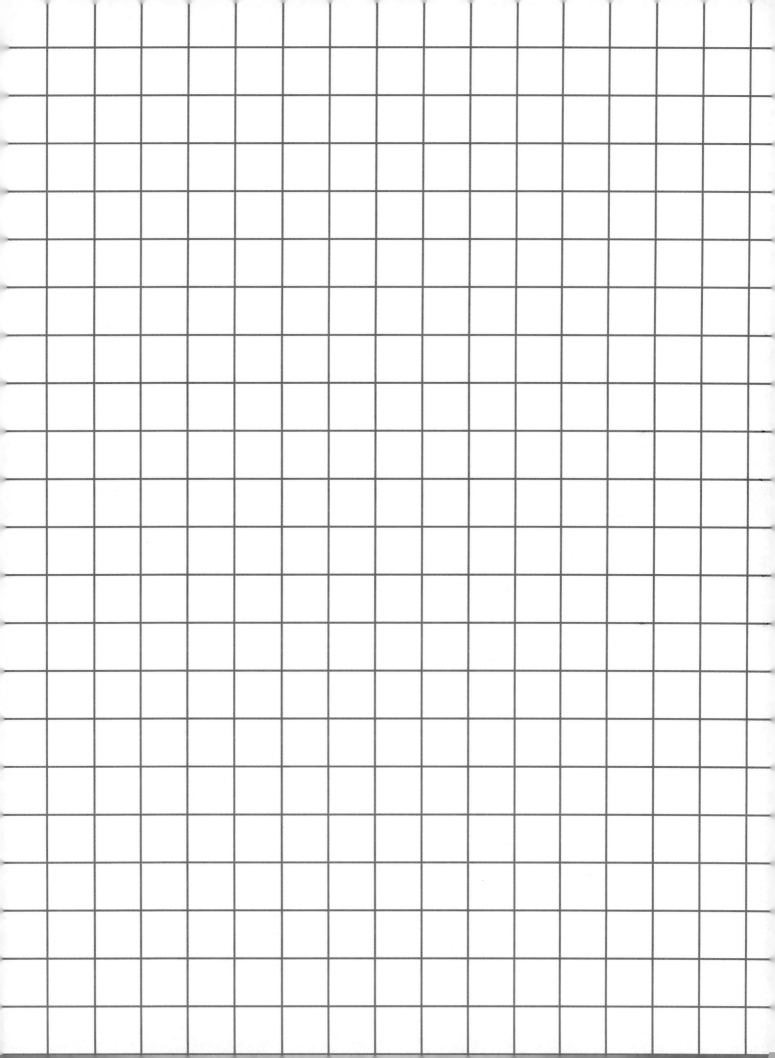

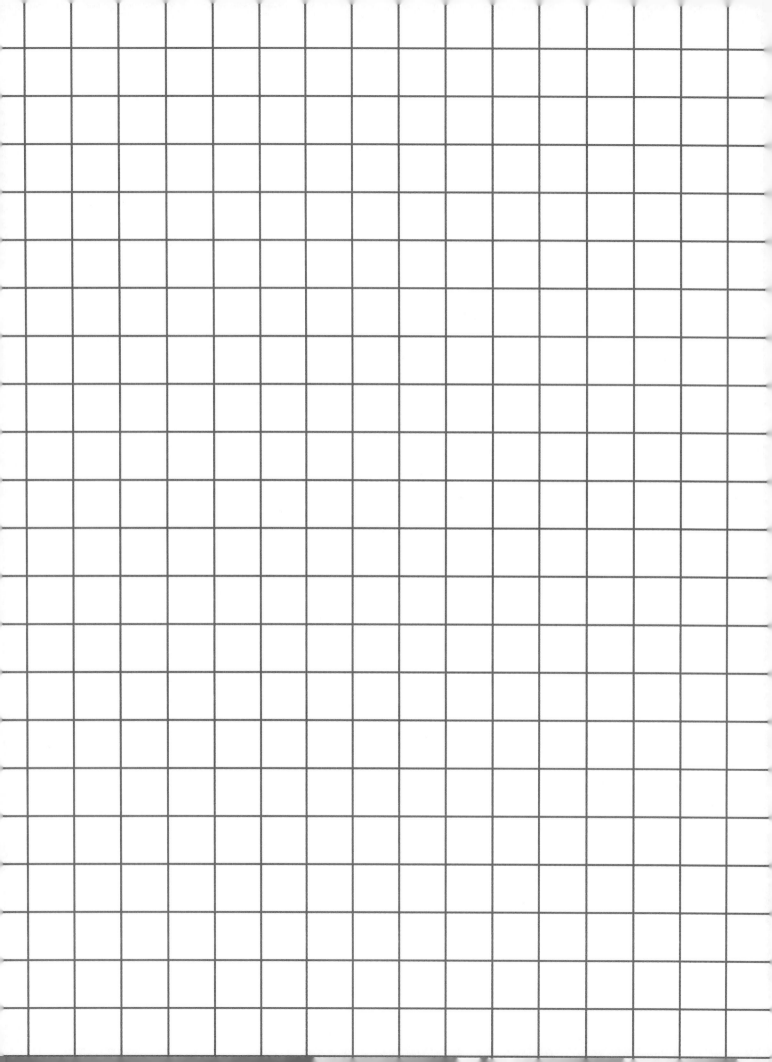

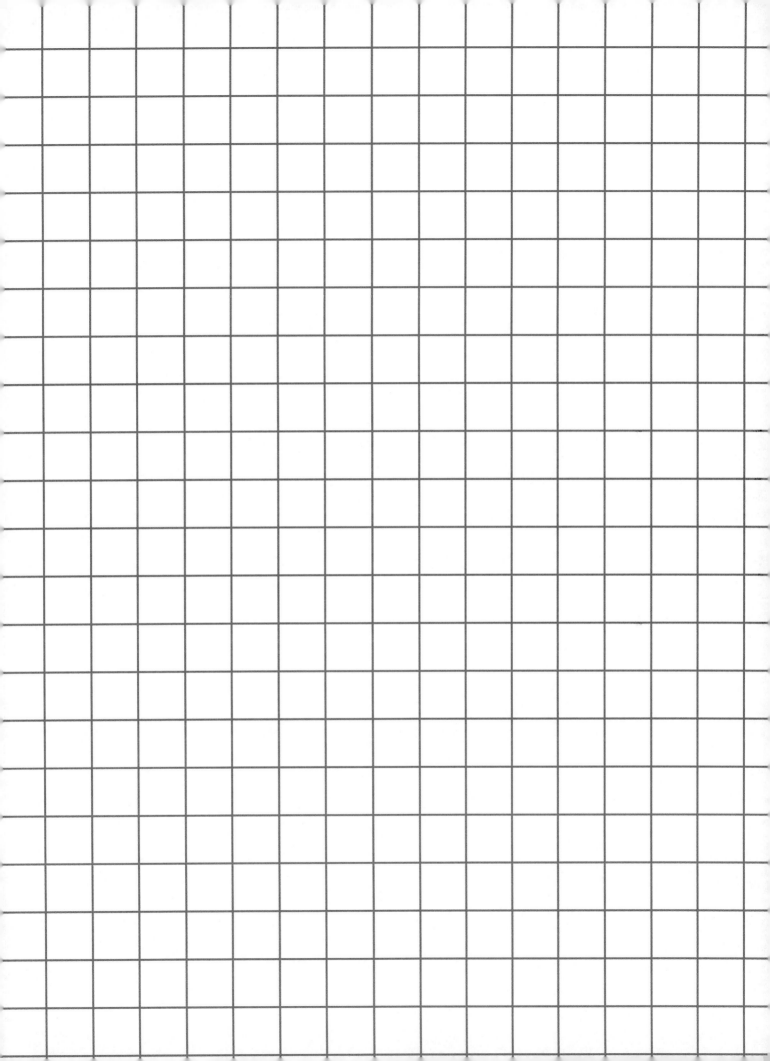

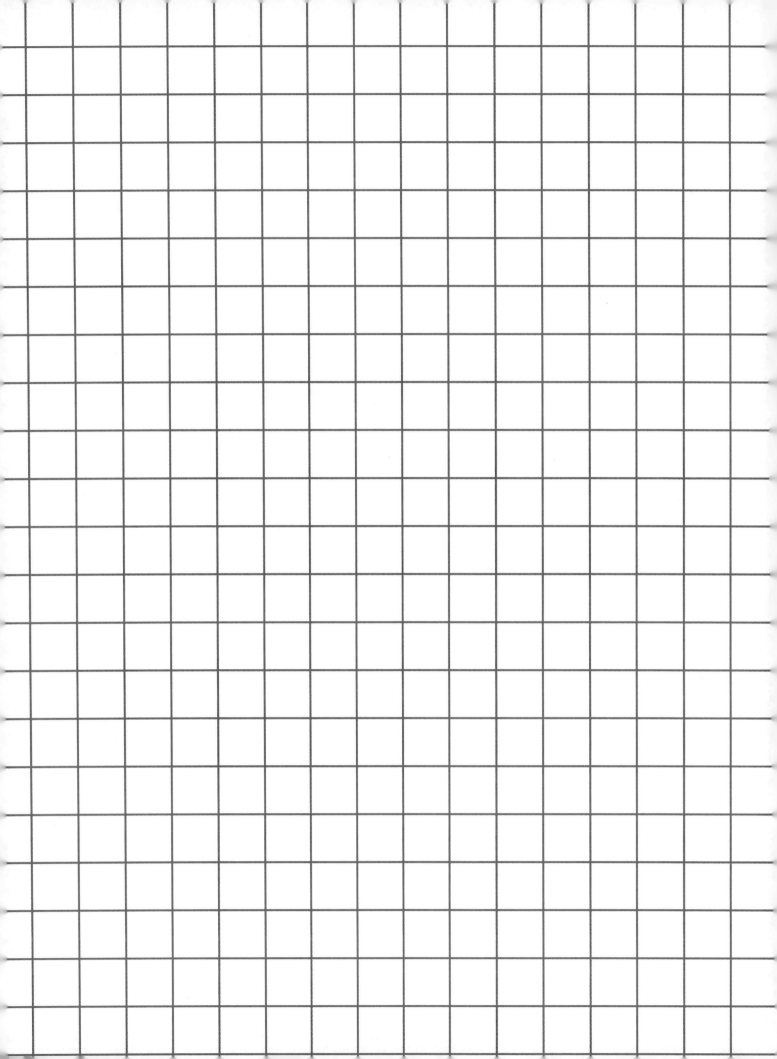

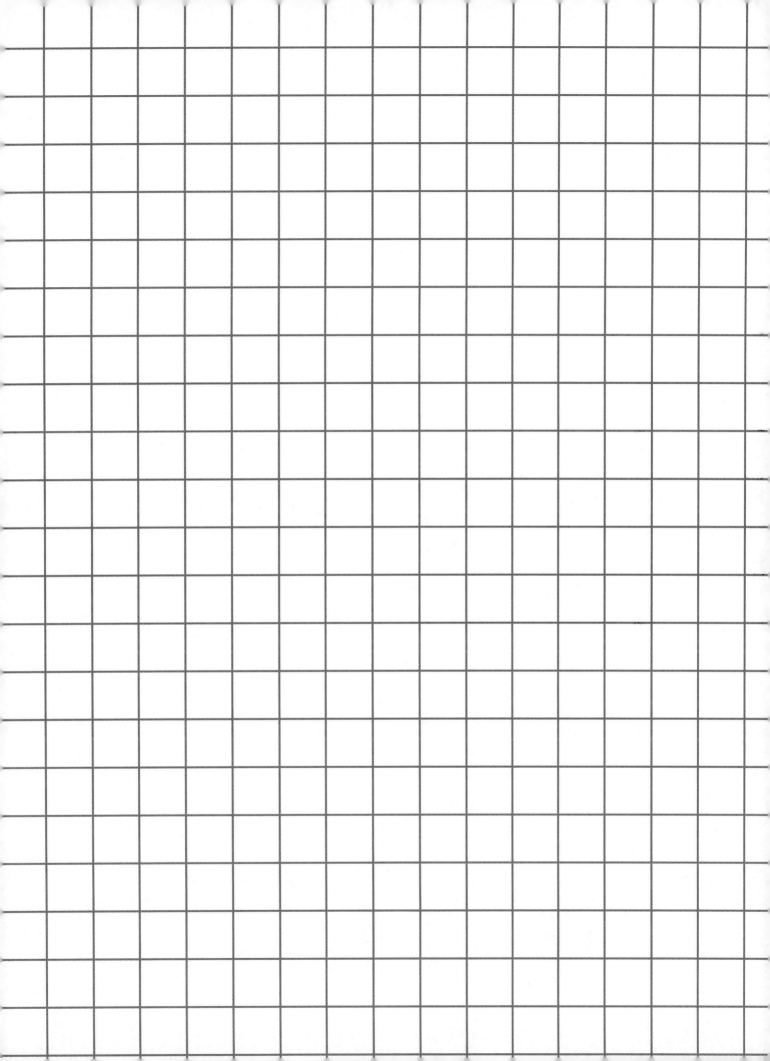

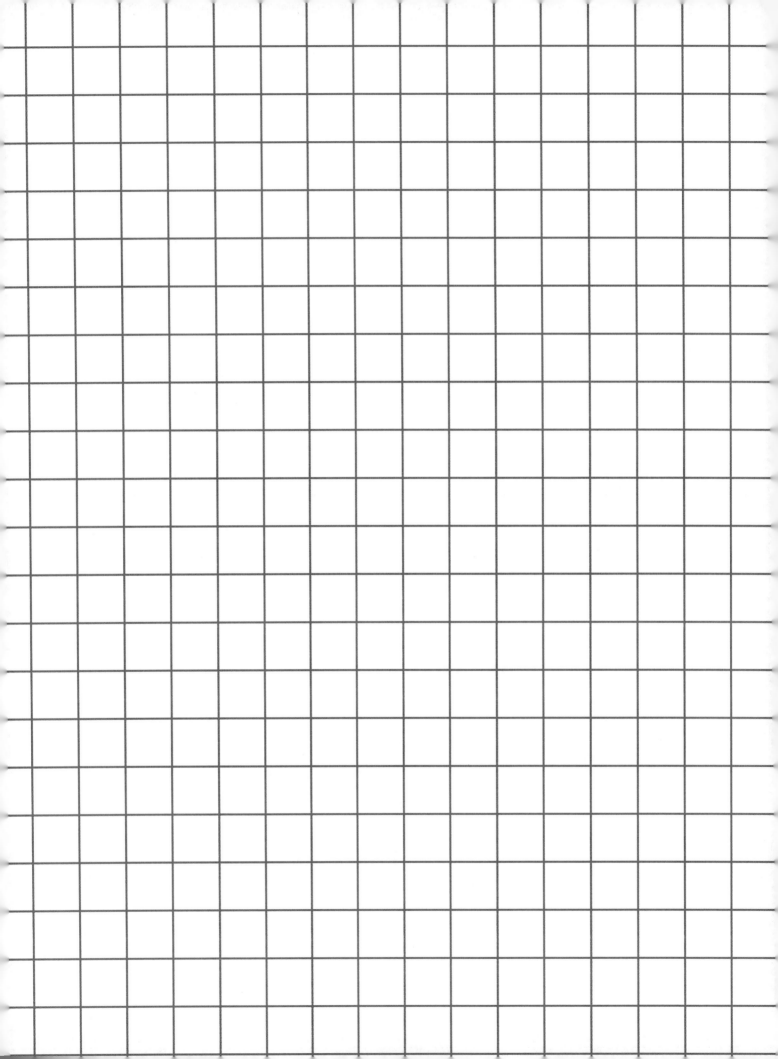

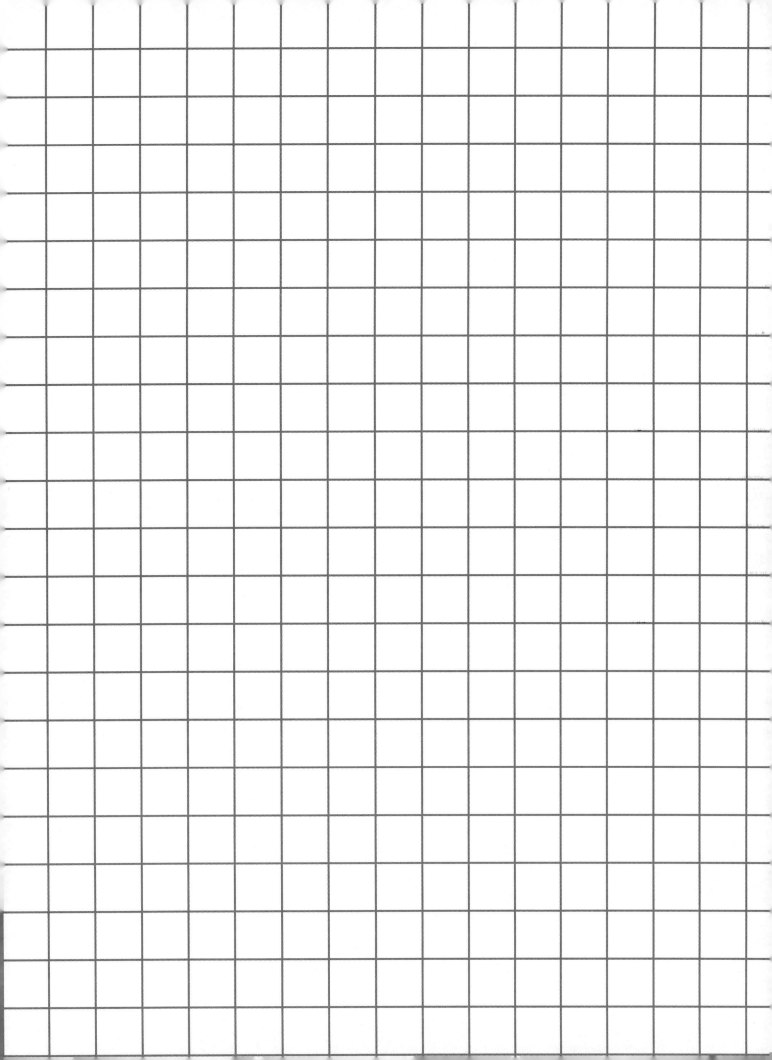

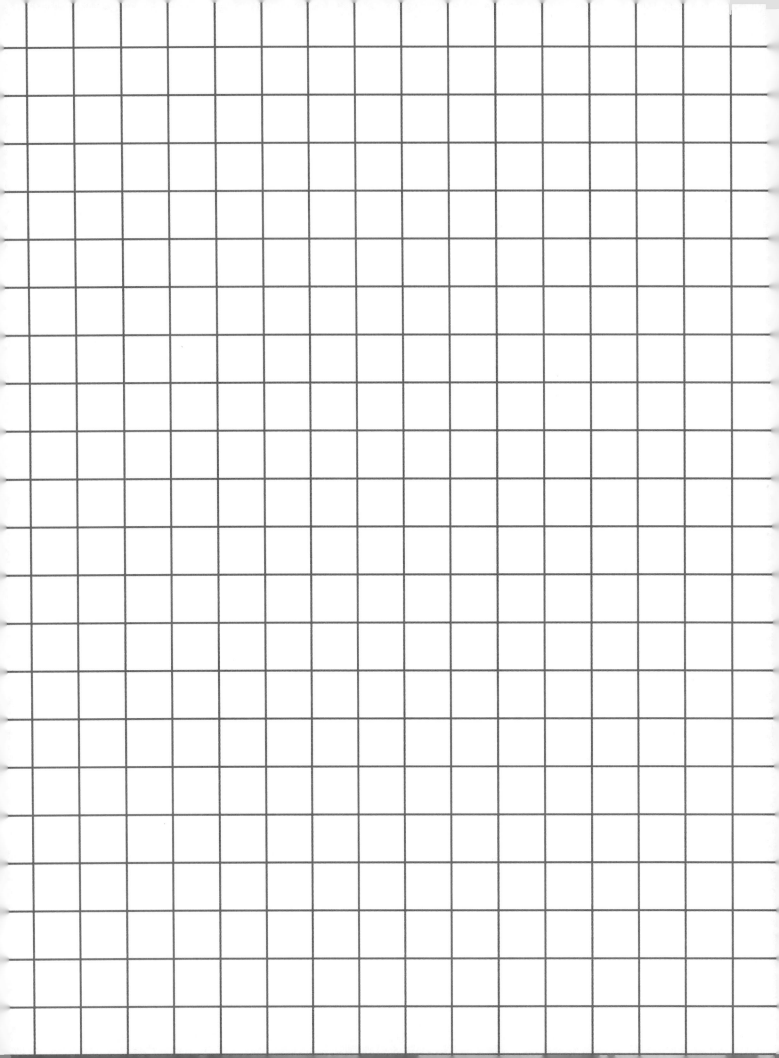

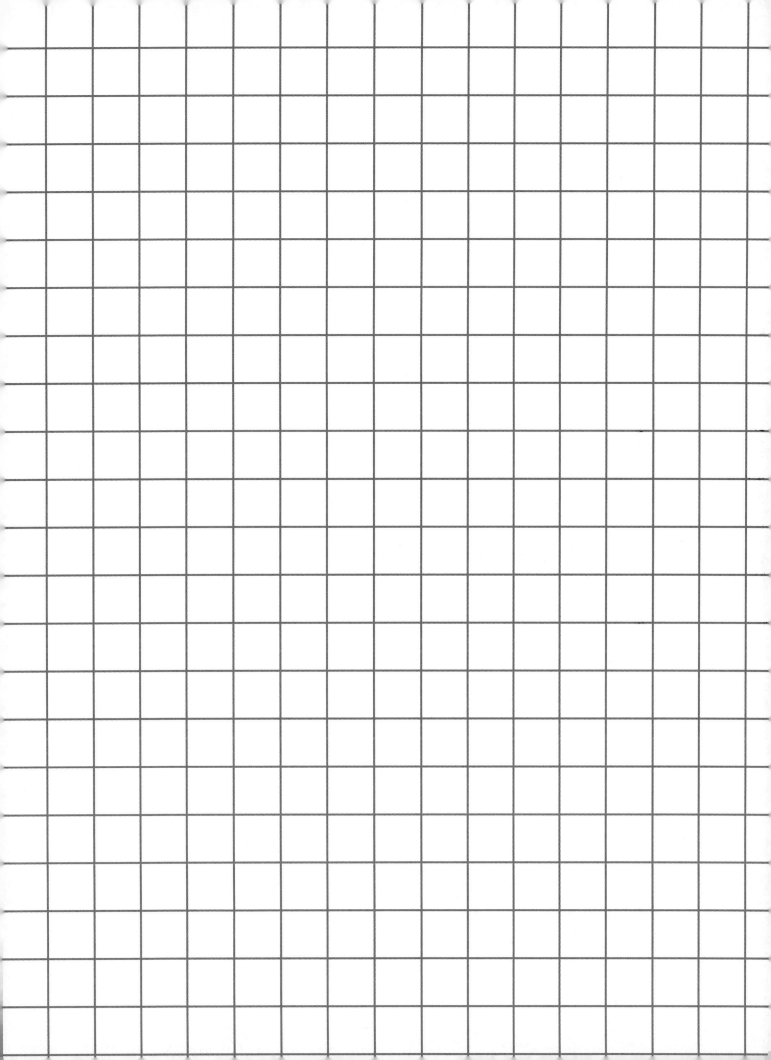

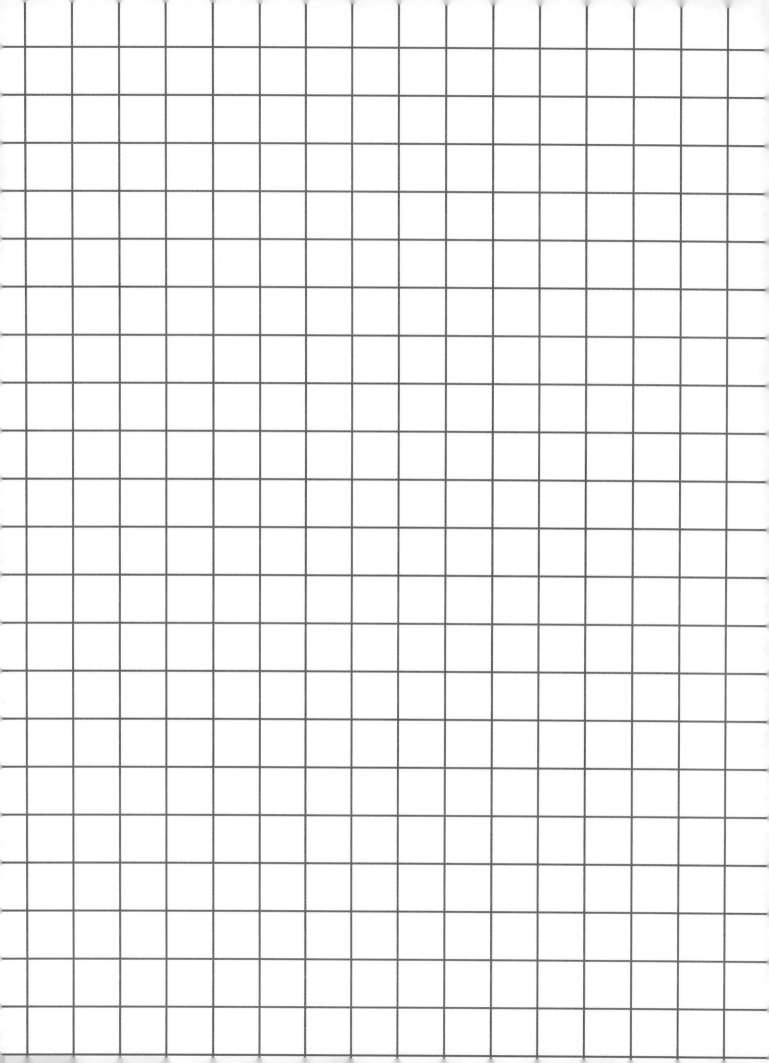

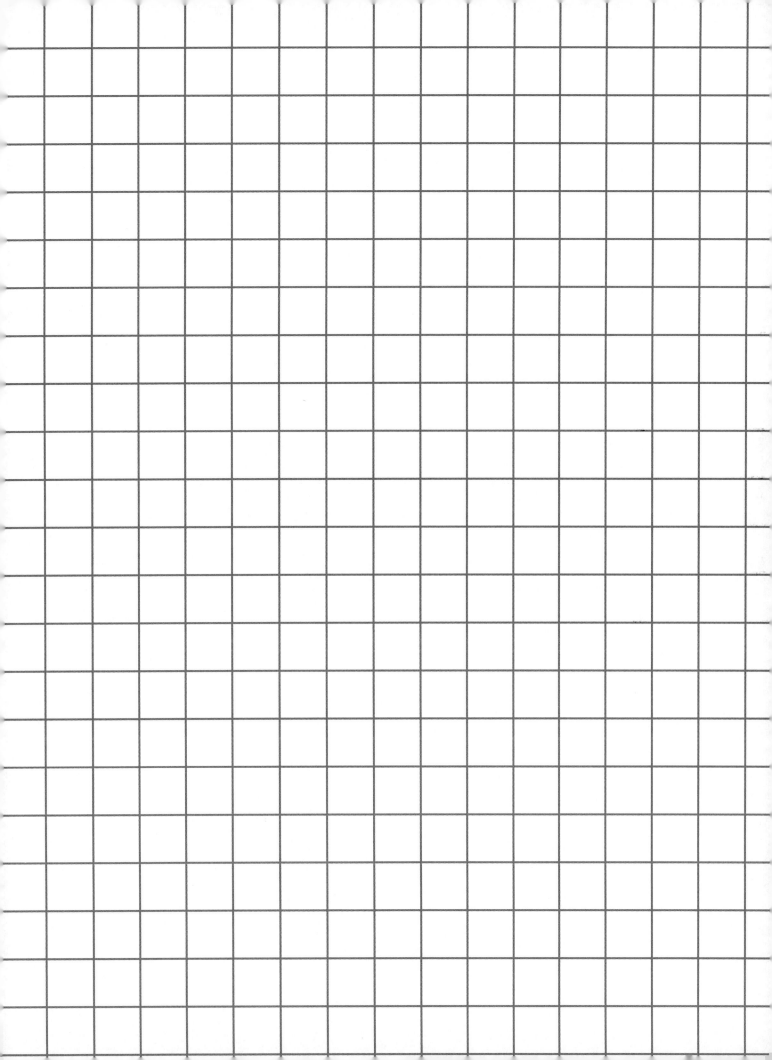

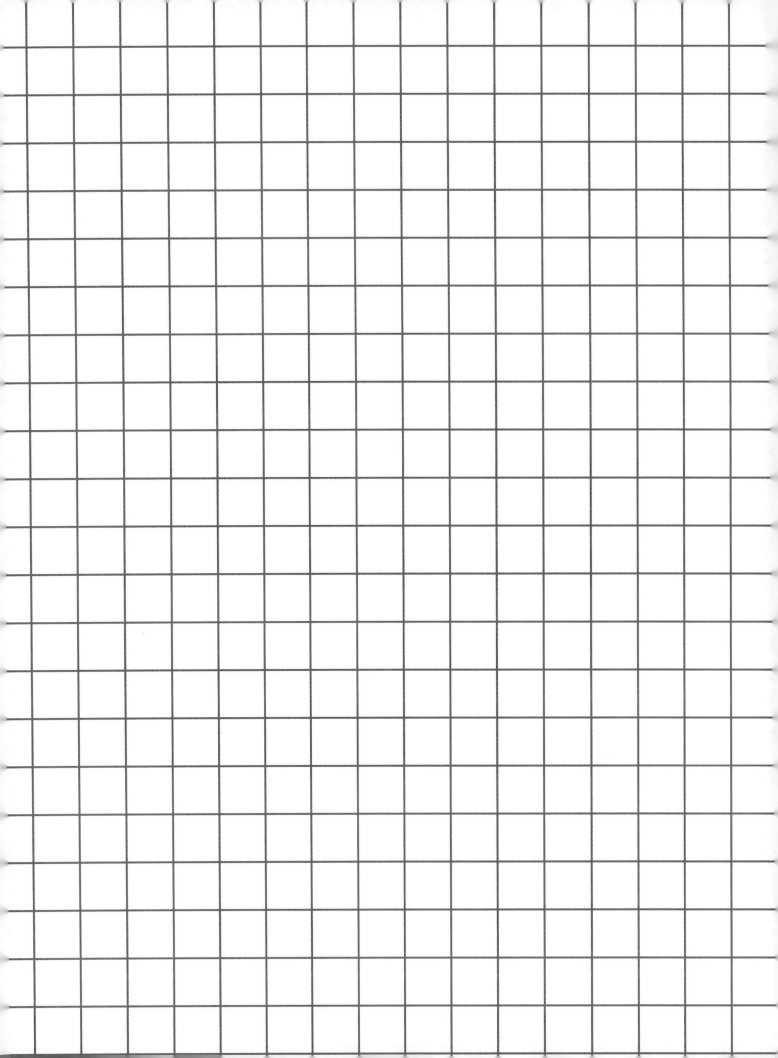

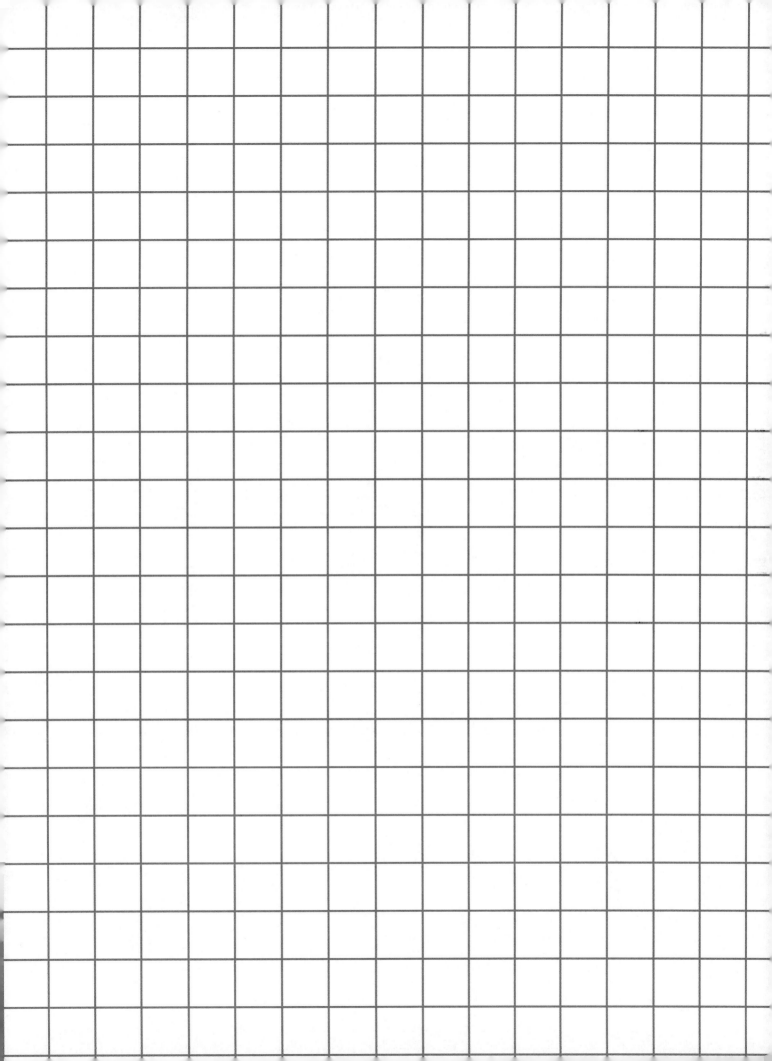

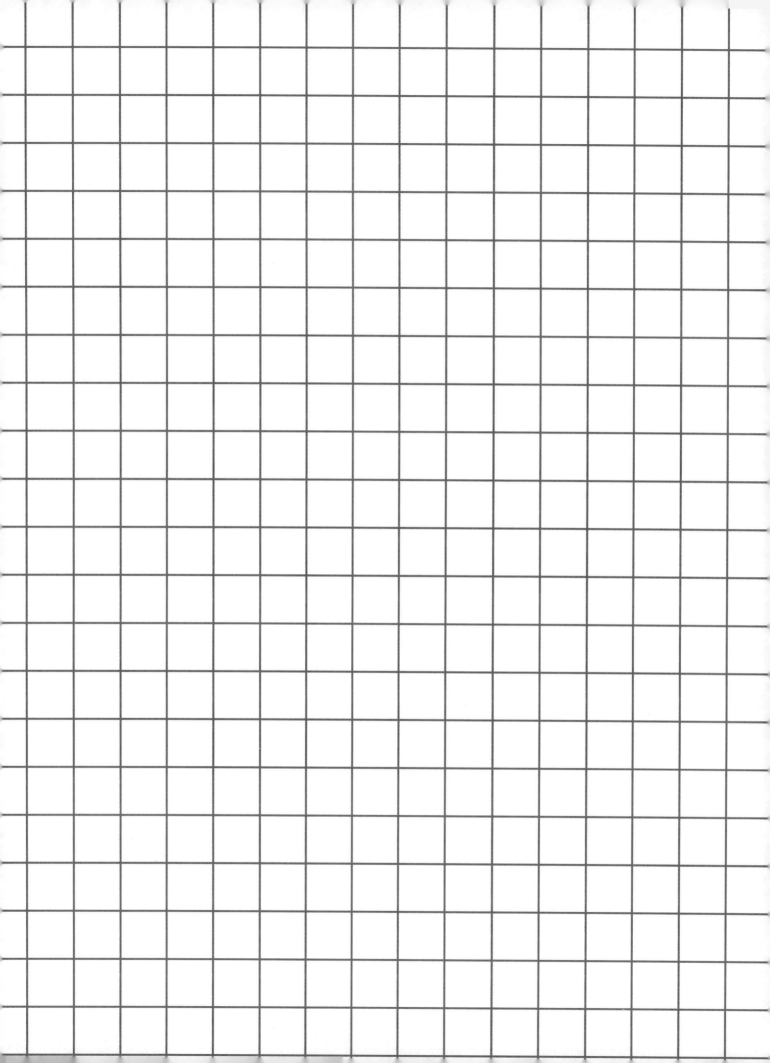

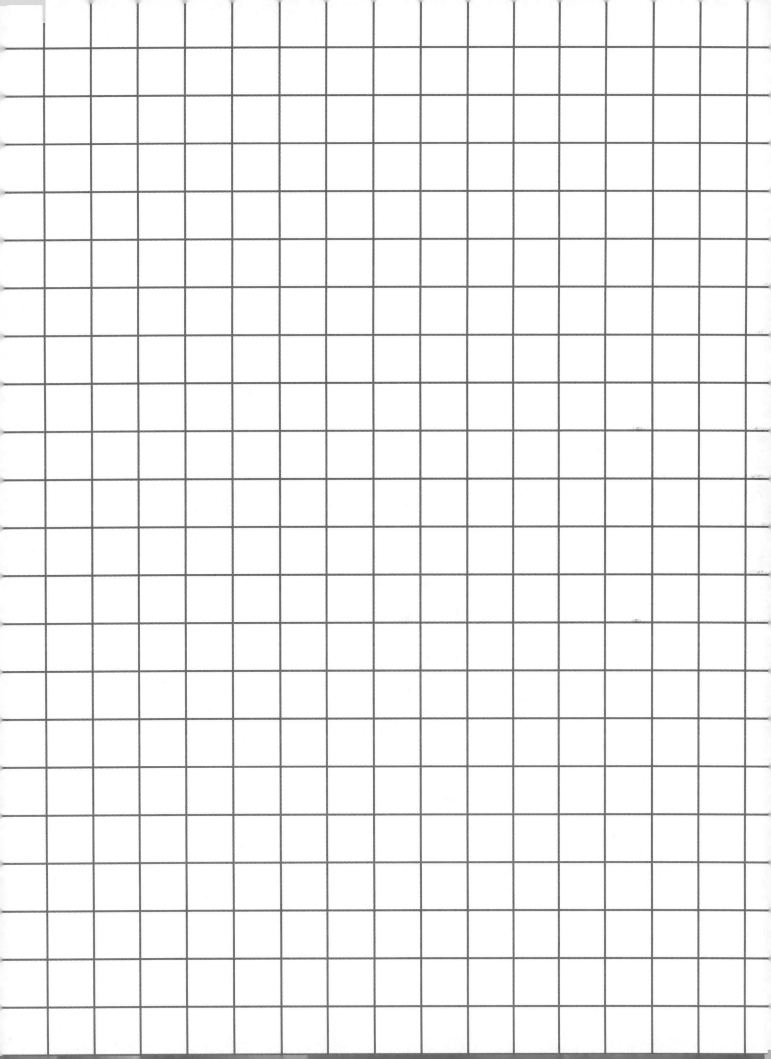

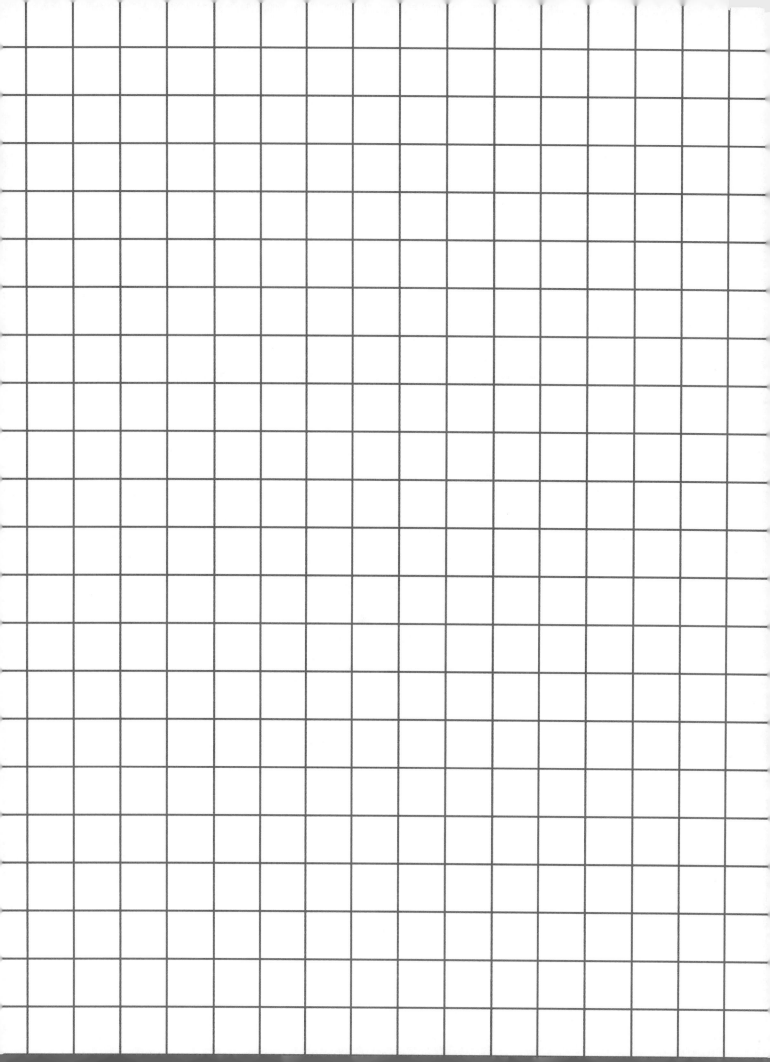

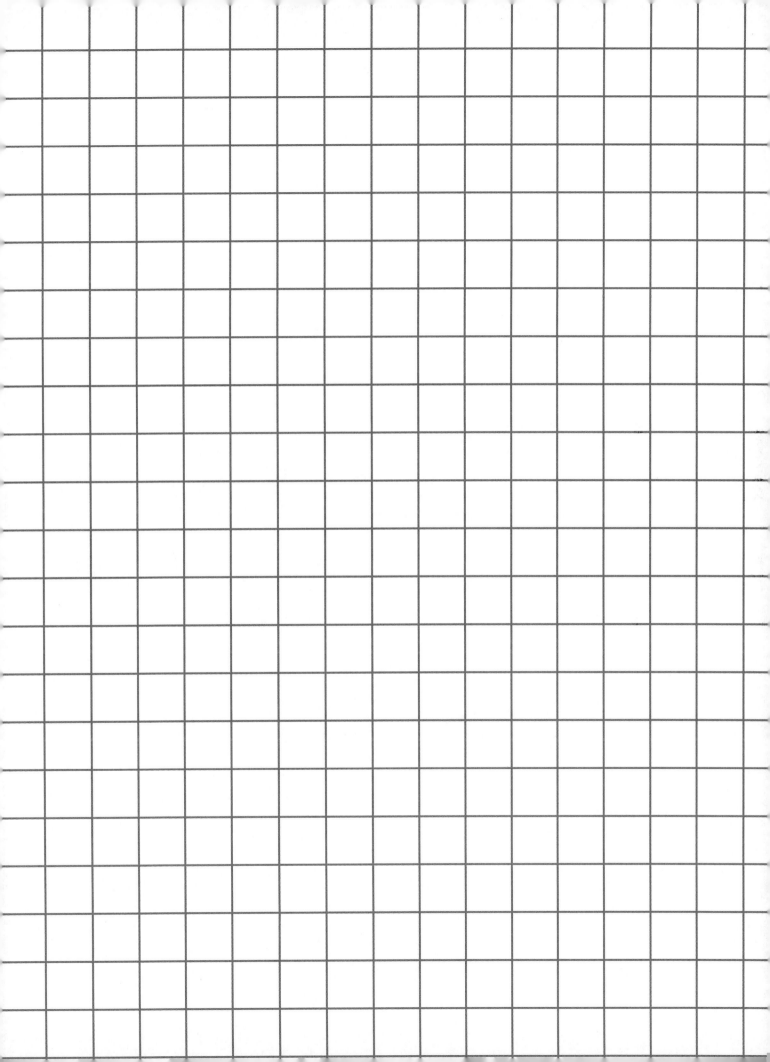

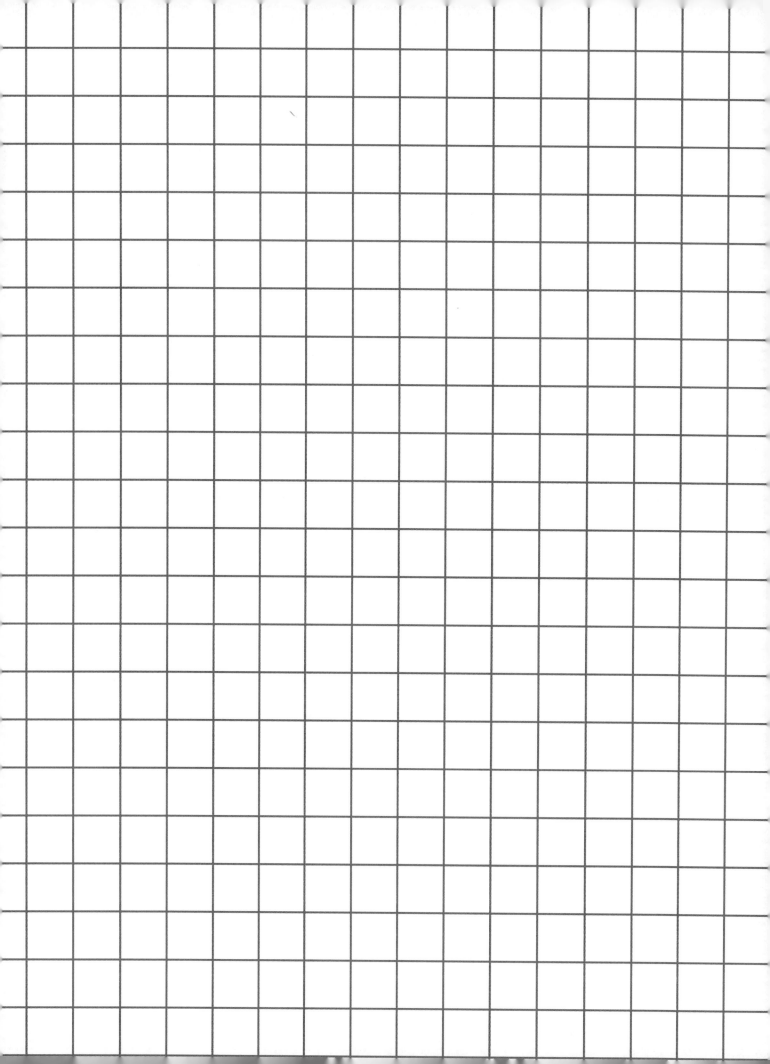

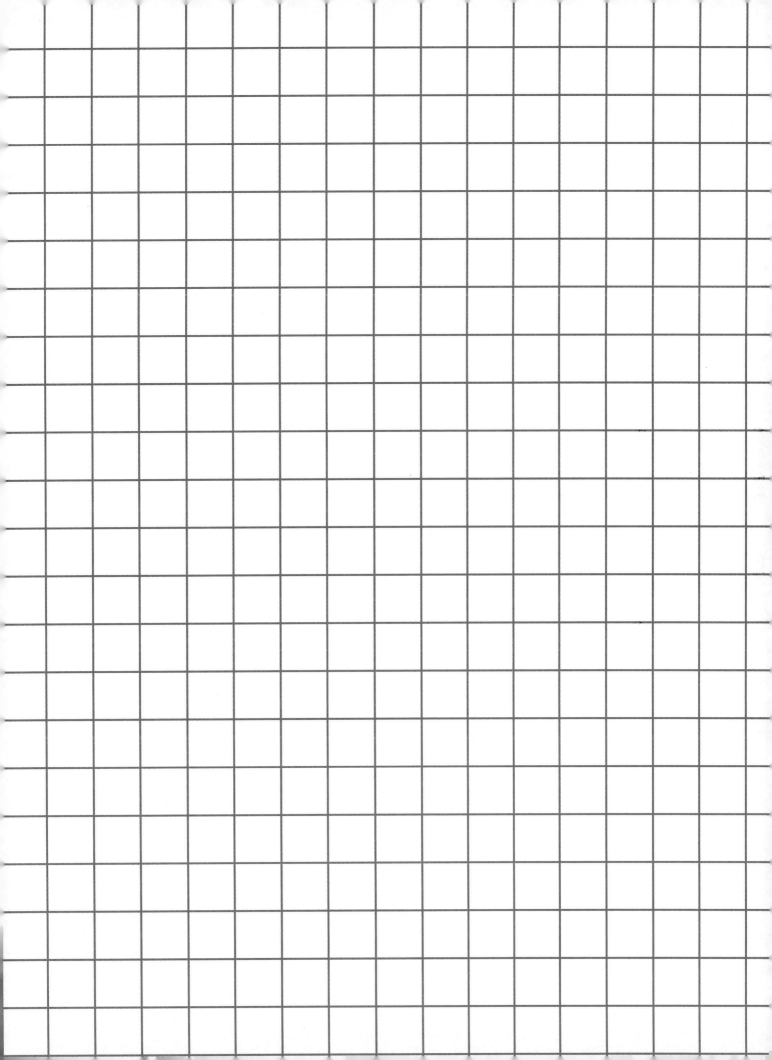

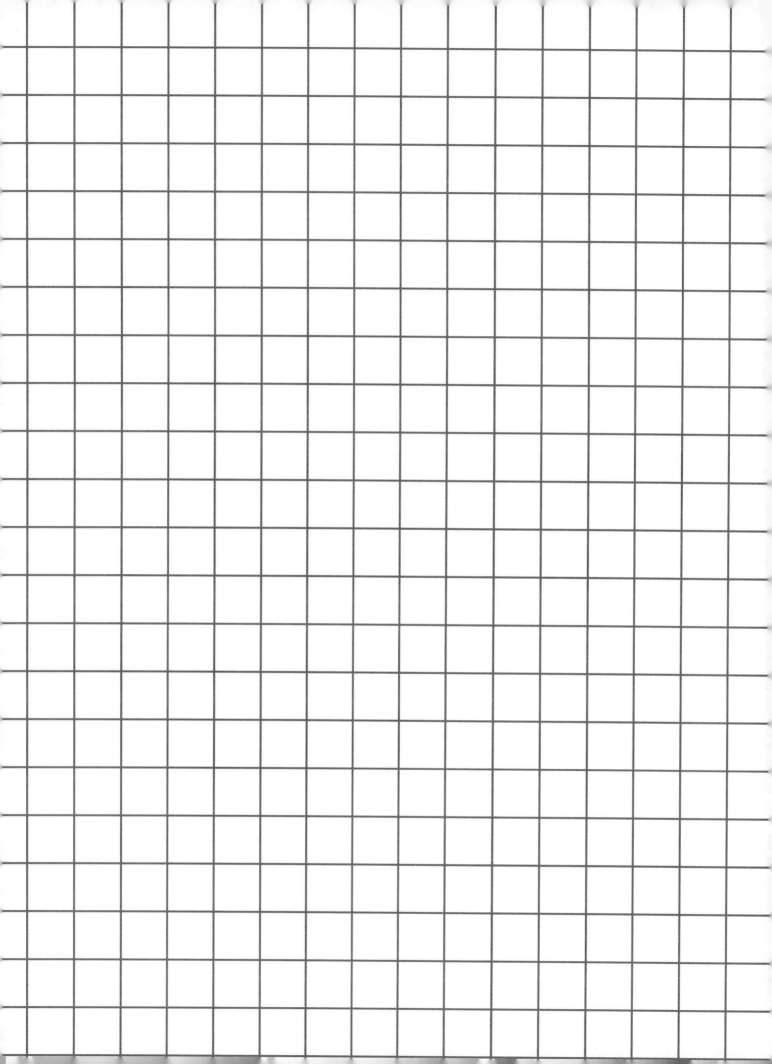

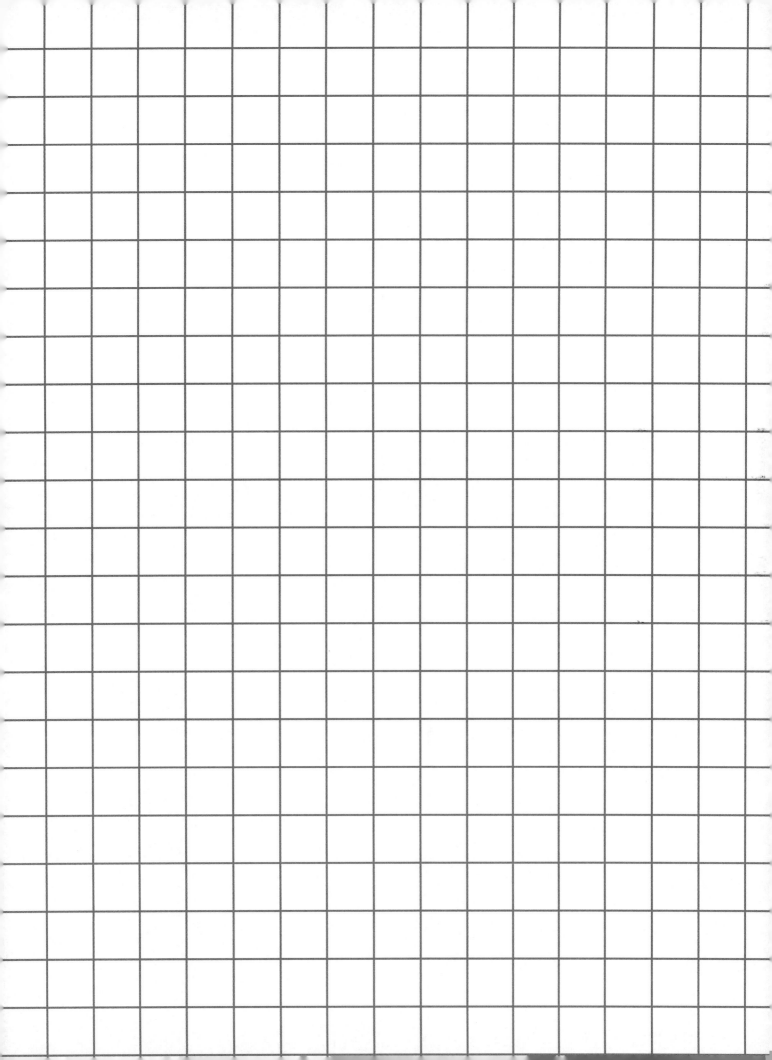

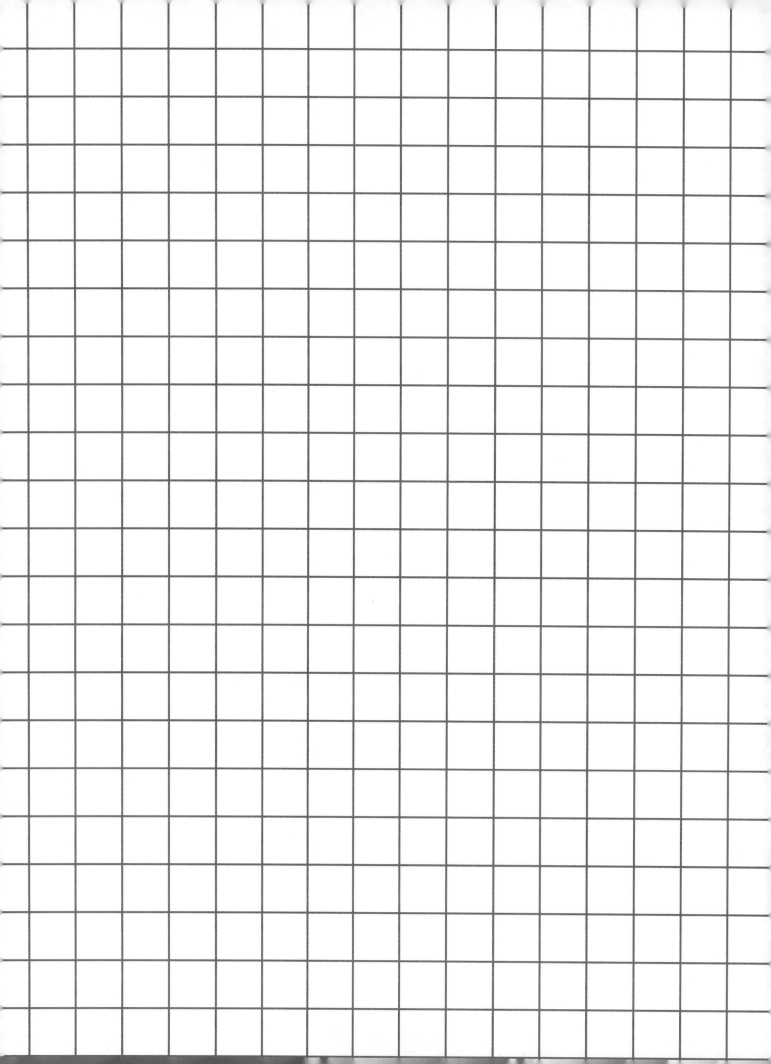

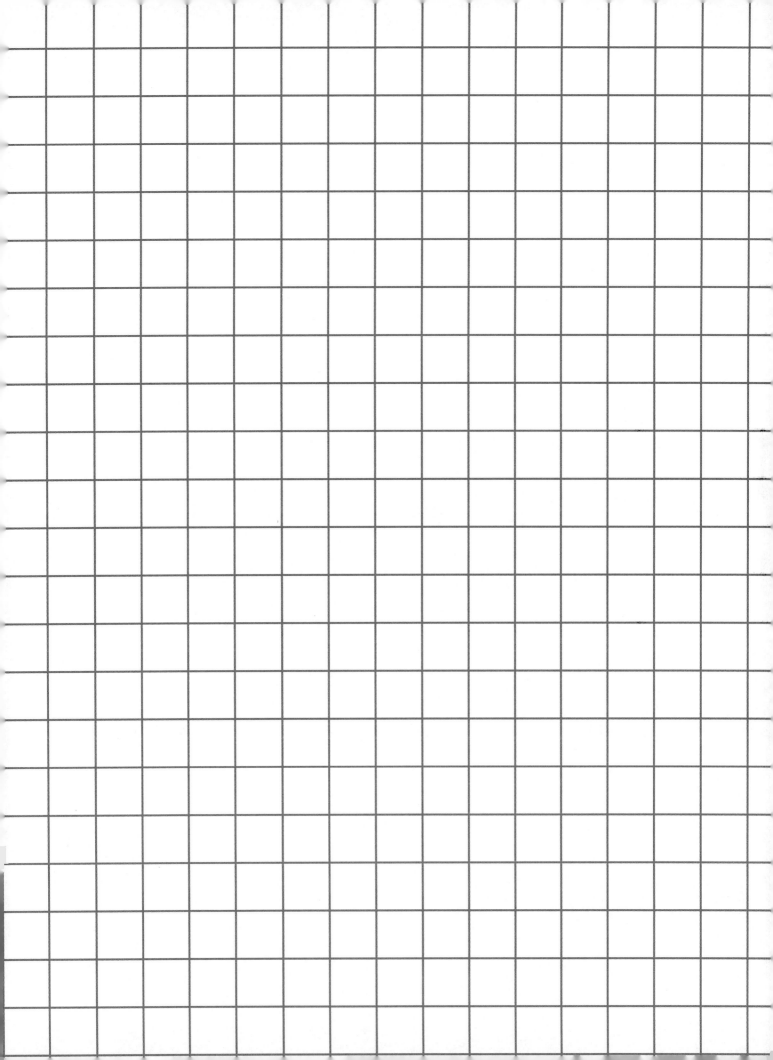

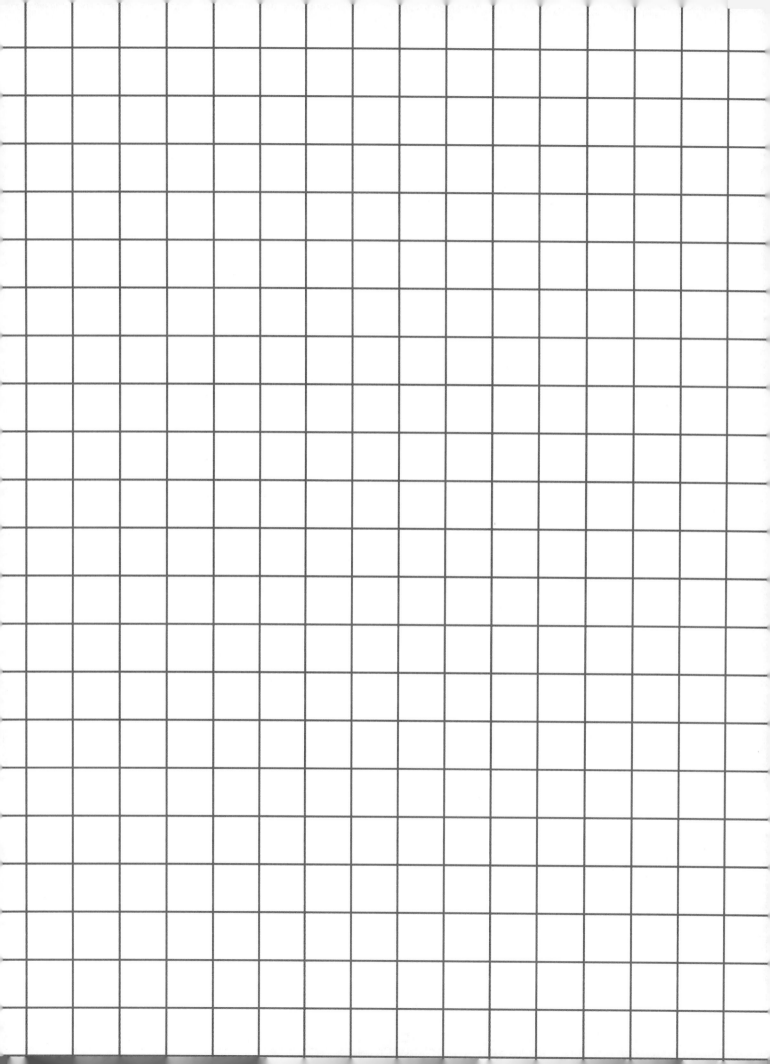

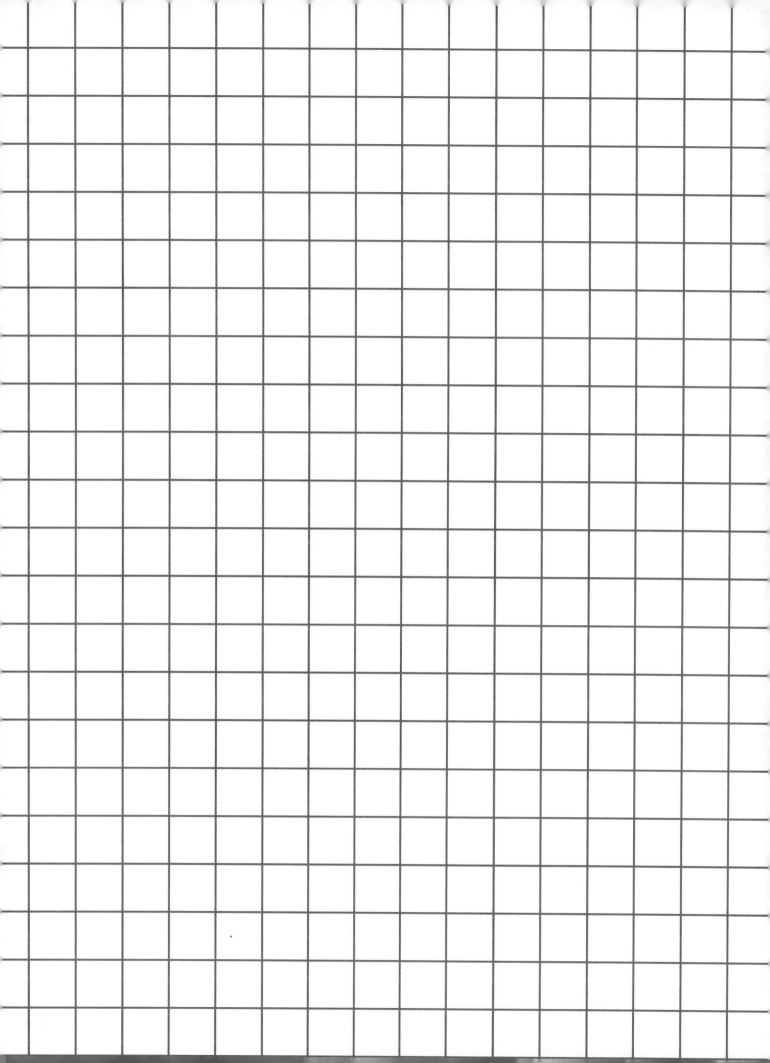

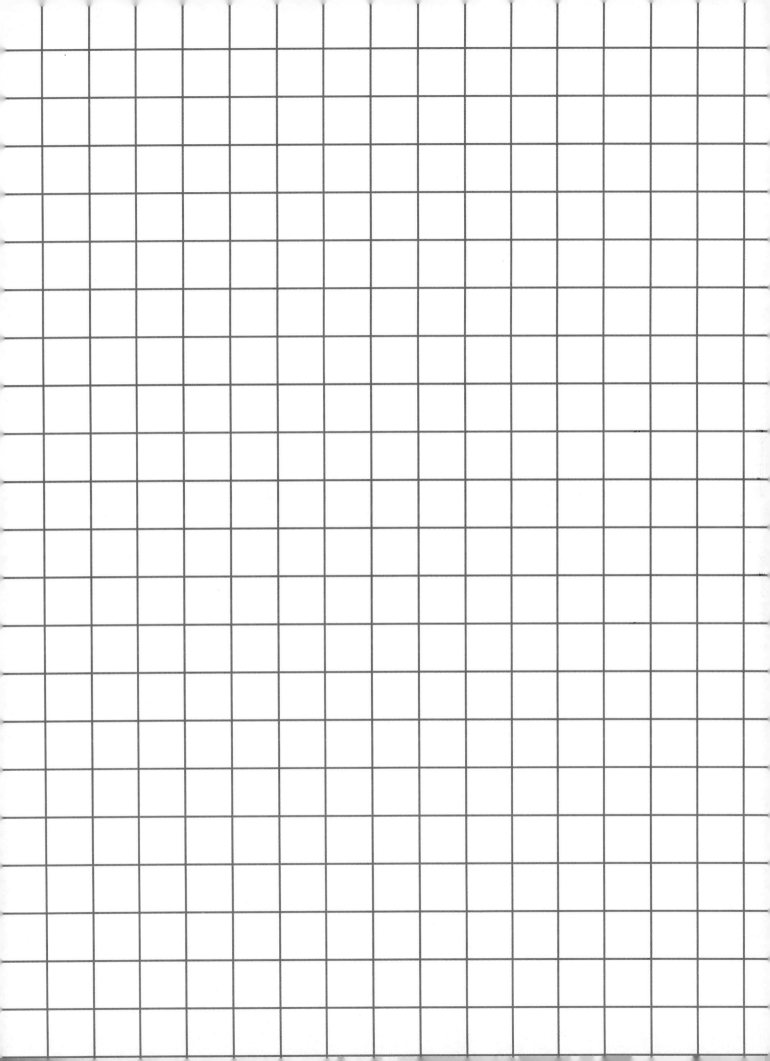

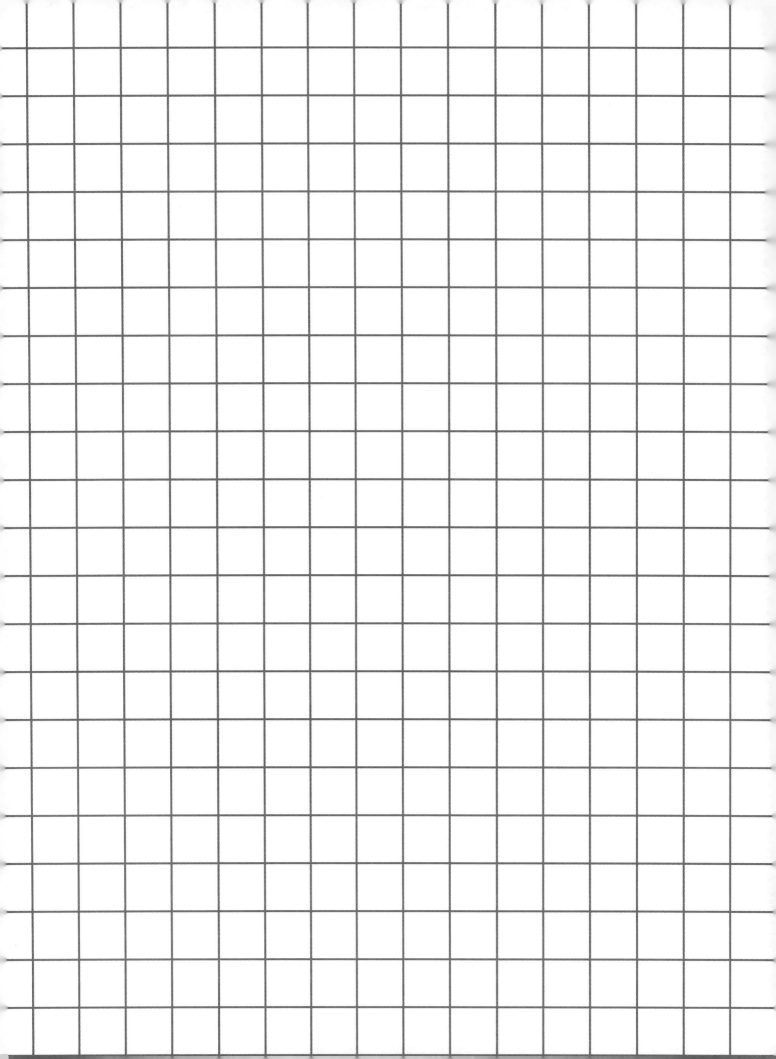

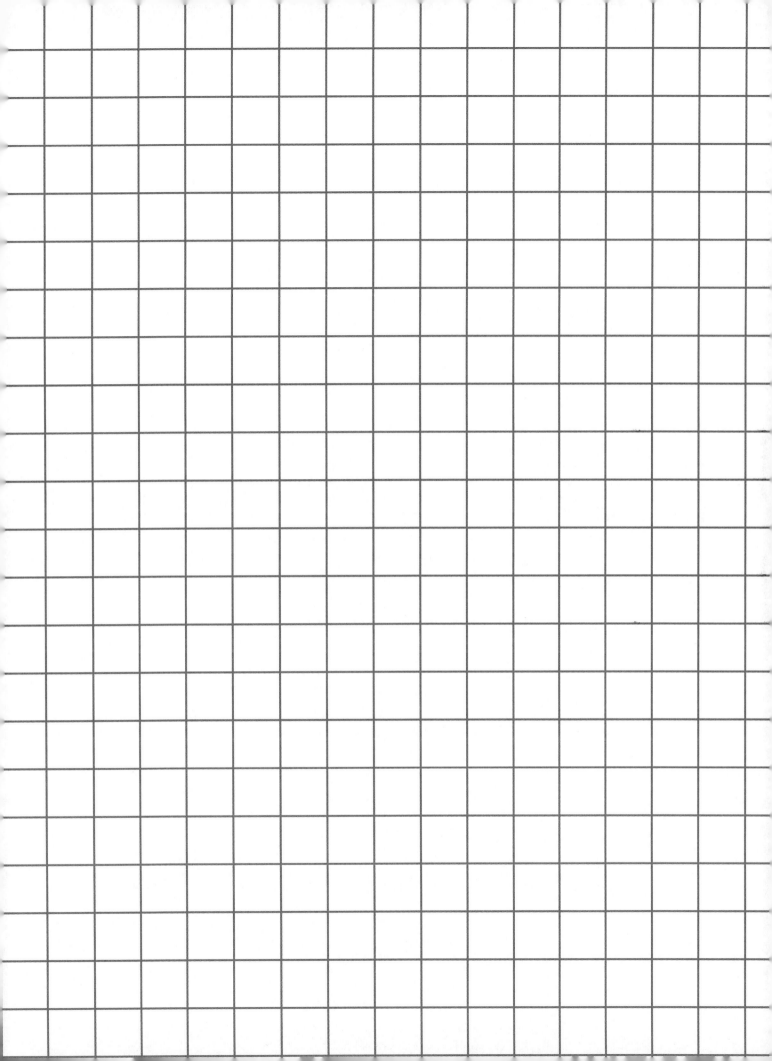

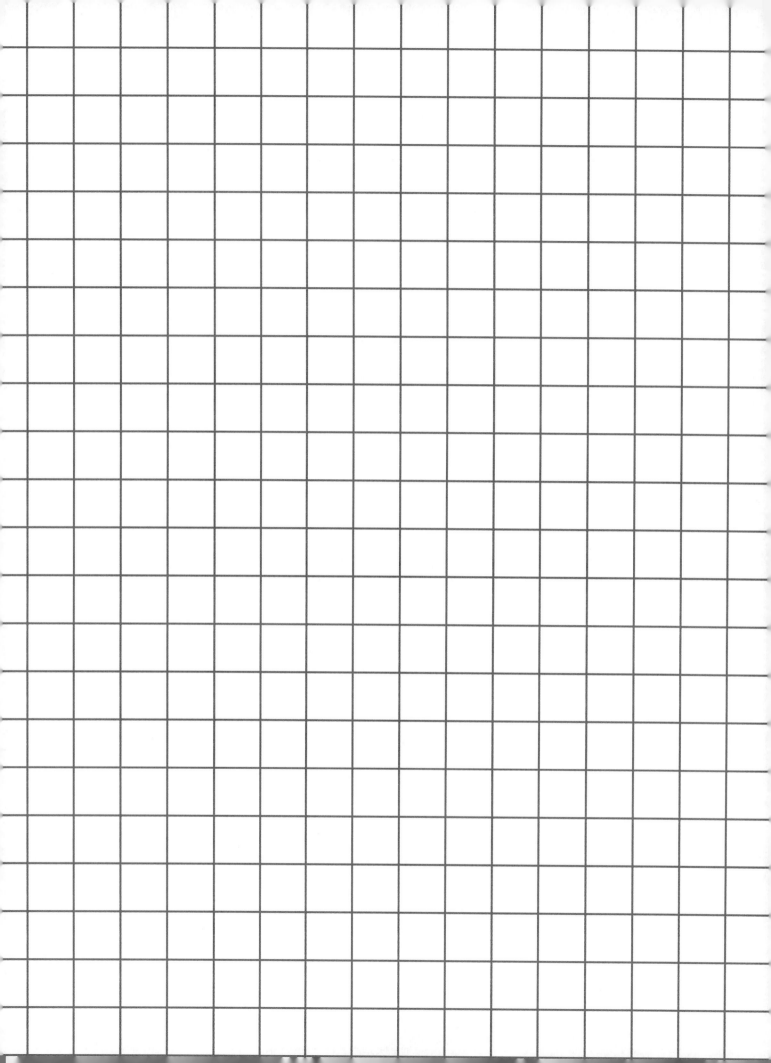

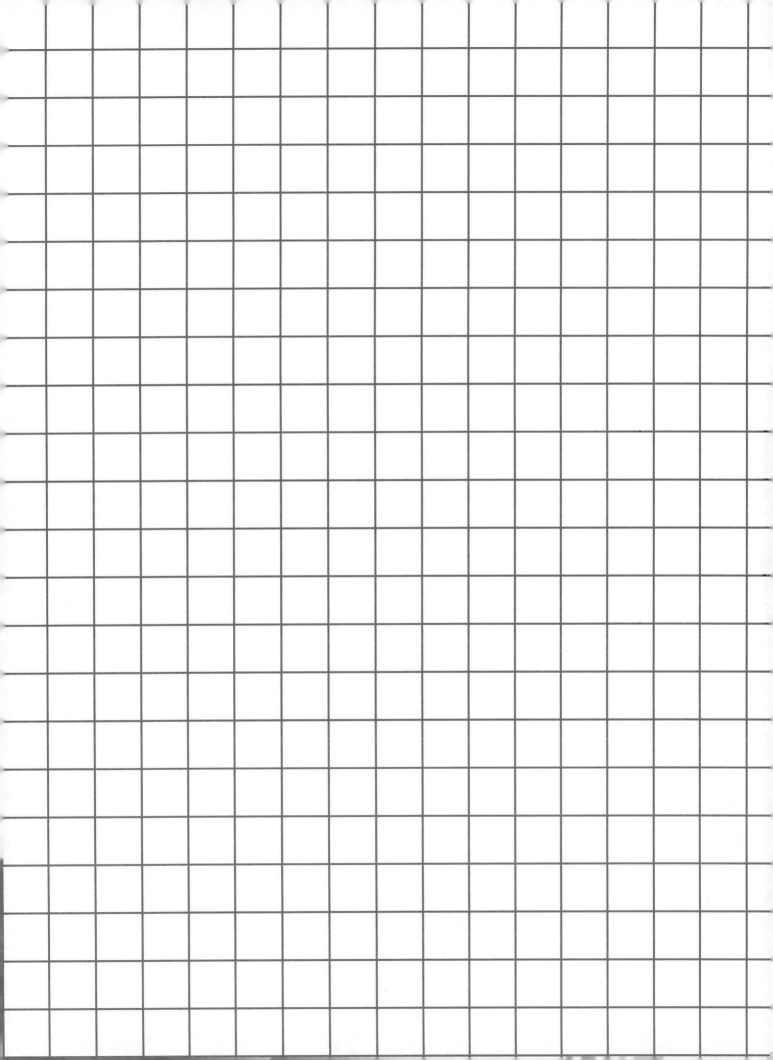

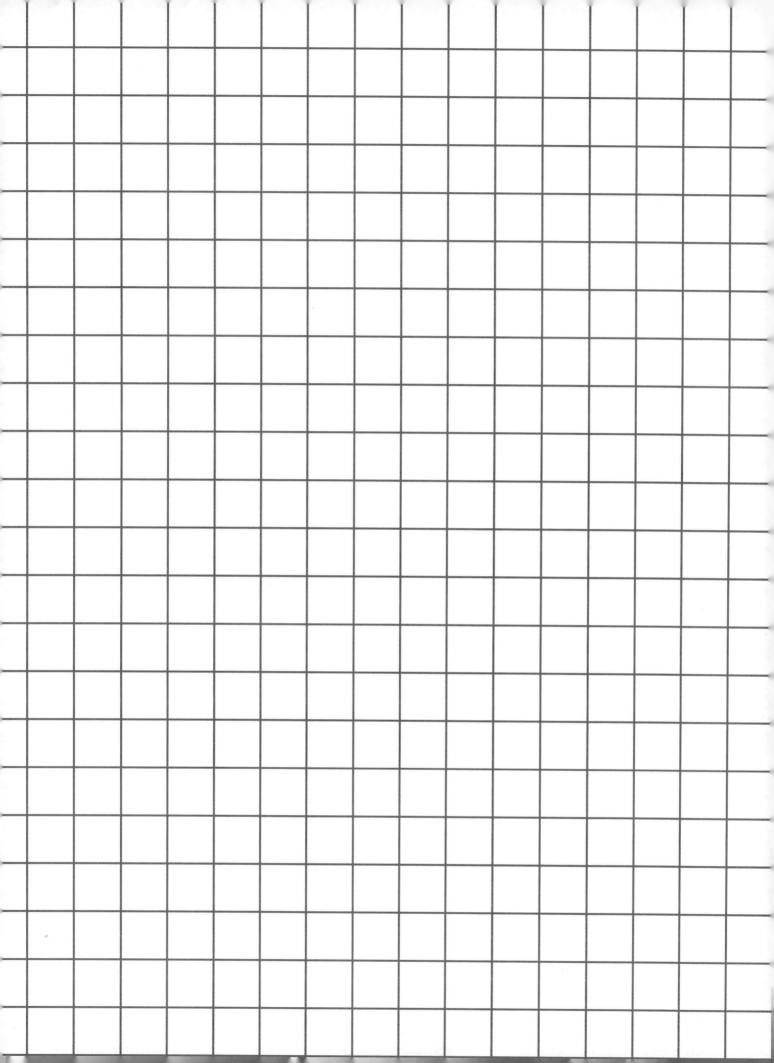

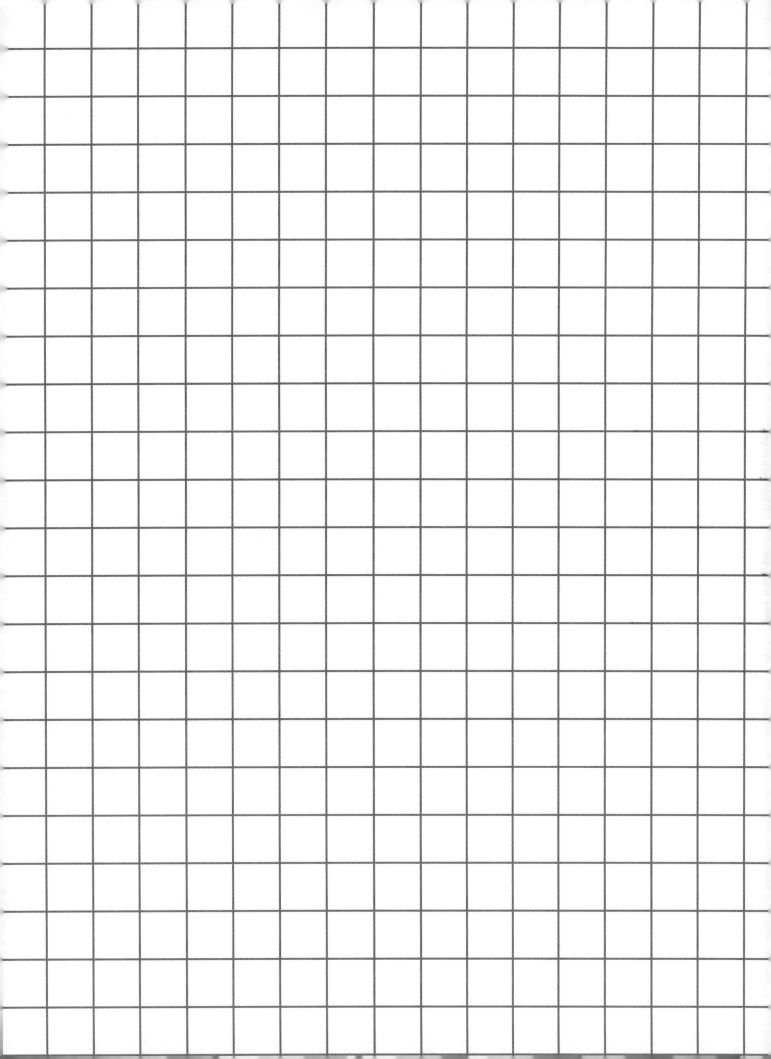

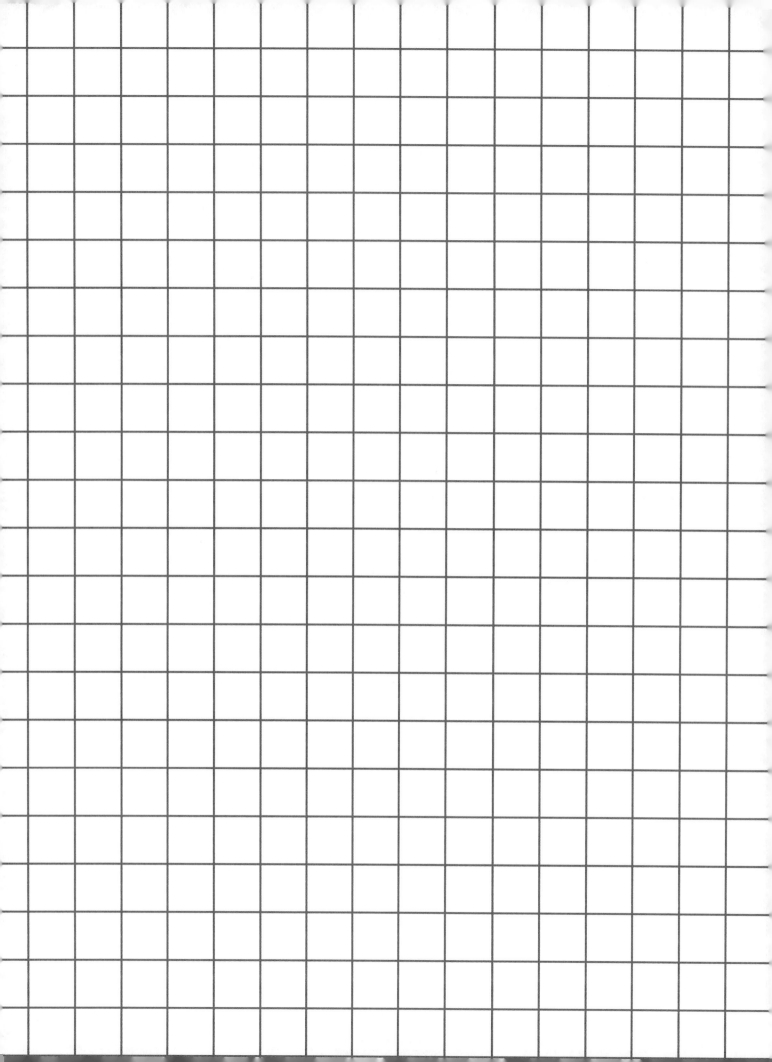

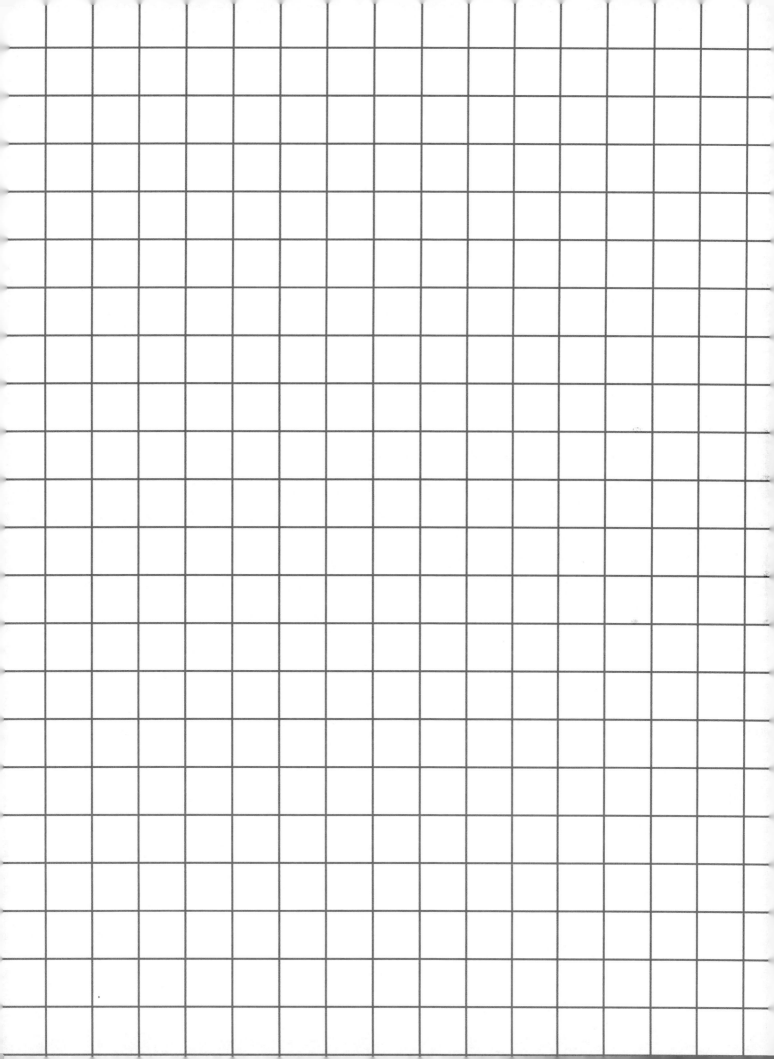

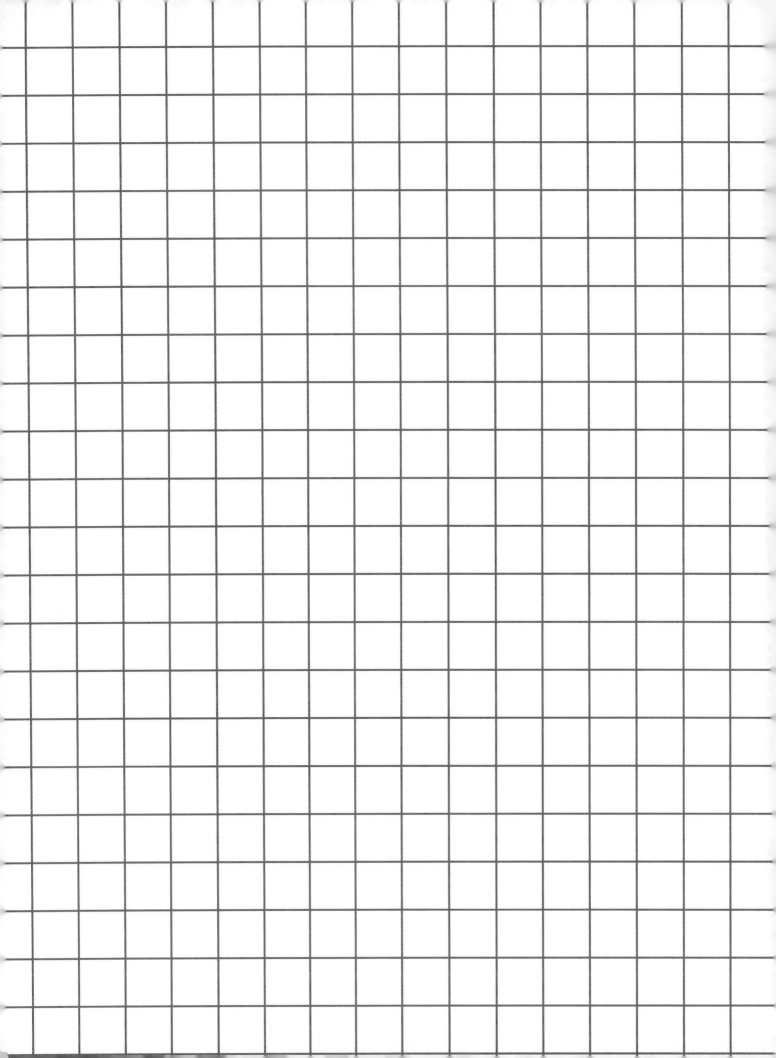

Made in the USA
Middletown, DE
29 August 2024

59958435R00071